People of the *Garden*
GARDEN OF THE GODS

People of the Garden
GARDEN OF THE GODS

Toni Hamill

East of the Mountains  and West of the Sun™

RHYOLITE PRESS LLC
Colorado Springs, Colorado

Copyright © 2024 Toni Hamill

All Rights Reserved. No portion of this book may be reproduced in any form or by any electronic or mechanical means, including information storage and retrieval systems, without permission from the publisher, except by a reviewer who may quote brief passages in a review.

Published in the United States of America by Rhyolite Press, LLC

P.O. Box 60144   Colorado Springs, Colorado 80960   www.rhyolitepress.com

People of the Garden Garden of the Gods, Colorado
Toni, Hamill

First Edition   February 1, 2024

ISBN: 978-1-943829-55-2

Library of Congress Control Number: 2024902969

Publisher's Cataloging-in-Publication data

Names: Hamill, Toni, author.

Title: People of the Garden : Garden of the Gods / Toni Hamill.

Description: Includes bibliographical references and index. | Colorado Springs, CO: Rhyolite Press LLC, 2024.

Identifiers: LCCN: 2024902969 | ISBN: 978-1-943829-55-2

Subjects: LCSH Garden of the Gods (Colorado Springs, Colo.)--History. | Colorado Springs (Colo.)--History. | Natural history--Colorado--Colorado Springs. | Formations (Geology)--Colorado--Colorado Springs. | Indians of North America--Colorado--Colorado Springs—History. | Sacred space--Colorado--Colorado Springs—History. | BISAC HISTORY / United States / State & Local / West (AK, CA, CO, HI, ID, MT, NV, UT, WY) | HISTORY / Indigenous / Creation & Origins | SCIENCE / Earth Sciences / Geology

Classification: LCC F784.C7 .H36 2024 | DDC 978.8/56--dc23

Cover Photograph: Harry L. Standley
Book design: Suzanne Schorsch   Cover design: Don Kallaus

To my husband Bill, for his encouraging me to complete my Masters of Arts Degree in American Studies, and his support for my seeking publication.

I also dedicate this book to the Department of American Studies, University of Colorado, Colorado Springs (UCCS), and to my Thesis Committee, Dr. Thomas Wynn, Dr. John C. Miller, Dr. Linda Watts, and Mr. William Arbogast, to thank them for their guidance and support.

Finally, to all my friends who pushed me to submit my manuscript for publication, thank you.

Cover Photograph by Harry Landis Standley (1881-1951). A Colorado Springs photographer specializing in the natural landscape. The majority of Standley's work was hand-tinted black and white photographs that he marketed from his north Tejon Street studio and gallery. He was a member of the Colorado Mountain Club and a founding member of Colorado Springs' AdAmAn Club, one of the "Frozen Five" and served as the club's photographer. He climbed all of Colorado's fourteeners long before it became popular, capturing high altitude vistas, beautiful mountain lakes, and enlightened the general public to Colorado's hidden wonders. Colorado Springs Pioneers Museum

Contents

Introduction	i
As Grasshoppers: Ute Legend	ii
Geologic Formation of the Garden of the Gods and the Springs at Manitou	1
Prehistoric Cultures	15
American Indians and the Garden of the Gods	27
Explorers, Trappers and Traders	53
Miners, Settlers and Entrepreneurs Pikes Peak Highway	65
The Garden of the Gods as a City Park	115
The Master Plan	137
The Future	149
About the Author	151
Bibliography	153
Index	161

Entry to the Garden of the Gods
Photo, Author's Collection, Kimberly Evans

Introduction

The impact of the Garden of the Gods struck me one day as I sat in an ancient rock shelter talking to my husband on my cellular phone. What an incredible history this place must have. After several years of research, this is the story of the Garden of the Gods that has emerged, beginning with the geological formation and ending with the master plan currently in place for its future use. The emphasis of this account is on the people who have interacted with the Garden from approximately 300 B.C. to the present. Glen Eyrie, Blair Athol, Red Rock Canyon, and the boiling springs at Manitou Springs are also included in this account because their use is closely tied to the use of the Garden of the Gods.

The Garden of the Gods, which is a city park located at latitude 105, longitude 39 degrees at an altitude of approximately 6,400 feet, has had special significance to the people who have come in contact with it since prehistoric times. This 1364-acre area in Colorado Springs, Colorado has been a seasonal camping site; a place of spiritual significance; a place of scientific investigation; a place of inspiration for artists, writers, and lovers; and an attraction for tourists and those who earn their living from tourists. It is also a place to learn about anthropology, archaeology, biology, botany, ecology, geology, philanthropy, sociology, and zoology. This is not just a place for scientists to study; it is a place which has, over the centuries, inspired writers and artists as well.

While researching this book, I came across the stories of a wide variety of people to whom the Garden of the Gods is a place of sacred or historical significance. Most of these people found the Garden a place of spiritual connection; some like Isabella Bird, author of *A Lady's Life in the Rocky Mountains,* found the Garden offensive at first, but gradually its uniqueness grew on them until they too were spiritually connected to the Garden of the Gods.

The Garden of the Gods has been a special place to many people throughout the ages, and, with proper management, will continue to be for the generations to come.

Grand Sentinel (Towers and Pinnacles), Photo, Colorado College

As Grasshoppers: Ute Indian Legend

For generations the chosen people had lived in peace, joy and contentment at the foot of their Holy Mount, now known as Pike's Peak. Each day as the sun arose, they lifted their souls in prayer to Manitou, the Supreme Being who, because of the love and devotion of his children, had stamped the image of his face on the Holy Mount. All the land of the country round about was the property of the chosen people from as far away as The Face of the Holy Mount could be seen. They were satisfied to stay within this area since they had all their needs supplied and were close to their beloved Manitou. They were weak in number and unskilled in the arts of war, depending for their safety on the love of Manitou, and they had always been sustained.

But suddenly, from out of the north came a fierce, barbaric host of giants, great and tall, accompanied by monstrous beasts that could devour the earth or tread it down. Resistance seemed hopeless, for before the invaders the chosen people were as grasshoppers, and so they fell back, seeking refuge within the shadow of the Holy Mount. Their foes swept down upon them with the speed of stampeding buffaloes and with the savage fierceness of the mountain lion. In their extremity, the chosen people prayed to Manitou, who saw their need and behold, caused a miracle! He was seen to turn his face and look upon the invaders and their beasts, who were immediately turned into stone. Whatever their location, attitude, or action, they were petrified in that instant and so formed the silent, ancient army of granite giants which may still be seen, no longer invaders but defenders, in the Garden of the Gods.

Geologic Formation of the Garden of the Gods and the Springs at Manitou

Three hundred-seventy million years ago a giant inland sea covered Colorado. Marine animals, especially shellfish, flourished.[1] The earth was restless. Forces beneath the American Continent developed to the breaking point. Great pressure beneath the earth's surface pushed massive blocks of rock through a weak part of the earth's crust, forming the Ancestral Rocky Mountains.[2] The peaks grew through slow, continuous uplifting for forty-five to fifty million years. These rugged new mountains reached from sea level to dramatic heights. Small streams flowed down the mountain sides, wearing them away even as the pressure in the earth was raising them higher, until the pressure ceased, and the movement stopped. The balance was destroyed. As the mountains slowly eroded, the streams widened into rivers. The force of the running water washed sand, gravel, and mud from the mountains, spreading the deposits in alluvial fans (gray-tan Fountain Formation) around the base of the mountains on to broad flood plains and deltas, and into the seas. Debris including vegetation such as the *Lepidodendron* or Scale Tree, settled to the bottom of the seas and compacted under the pressure of the water into sedimentary rock thousands of feet thick. Groundwater minerals flowed into the sediment creating the deep red Pennsylvanian and the pinker Permian layers.[3] Over the next forty million years, the Ancient Rockies gradually disappeared. Their peaks leveled and the sea withdrew.[4] Sand dunes (white to red Lyons Sandstone, depending on the amount of iron oxide present) remained where the sea had withdrawn.[5]

During the Jurassic Period two-hundred million years ago, the climate changed and moist tropical air warmed Colorado. Fresh-water streams flowed into lakes and swamps.[6] In the humid lowland climate, Stegosaurus grazed off horsetails, cycads, conifers, and seed ferns. Fish-like reptiles, crocodiles, ammonites, sharks, and rays inhabited the lakes and lagoons.[7]

One hundred to one hundred twenty-five million years passed. A quiet shallow Cretaceous Sea up to one hundred feet deep spread from the Arctic to the Gulf of Mexico.

Tylosuarus (*mosaur*),[8] thirty-three- to forty-foot-long marine lizards, ambushed their prey of fish, squid, birds, turtles, and other *mosasaurs*. These prey animals swam in the sea and crawled on the fine gray mud of the bottom. When they died, their shells or skeletons or teeth were buried in the mud and preserved as fossils in the fine, gray Pierre Shale. On the shore surrounding the sea, *Iguanodons* including *Theiophytalia kerri*[9] and *Ankylosaurus* browsed on the primitive flowering plants called *angiosperms* and on ferns, horsetails, and trees like conifers and gingkos.[10]

Seventy-five million years ago the Cretaceous Sea slowly withdrew and left the fine gray mud of Pierre Shale and sand behind.[11] Mountains began to form, raising the buried Fountain and Lyons formations and other sedimentary rocks to vertical positions along a network of faults, separating the mountains from Colorado Springs. The Rampart Range Fault runs north and south and borders the mountain uplift north of Fountain Creek. The Ute Pass Fault runs northwest from the valley of Fountain Creek into the mountains to Woodland Park and separates Cheyenne Mountain and Pikes Peak (14,110-foot elevation) from the Rampart Range (9,000-foot elevation). The Garden of the Gods, Glen Eyrie, and Red Rock Canyon are wedged between the Rampart Range Fault and the Ute Pass Fault.[12]

Aerial View (013-1083), Pikes Peak Library District, Stewart

As the mountains rose, basins sank to the east and west. Swift rivers and streams carved into the hills, filling the sinking valleys with gravel, sand, and silt. For twenty million years this rising, sinking, and filling continued. Pressure beneath the surface of the Rocky Mountains was building. Finally around twenty-eight million years ago the stress and strain was so great that the entire area from mid-Kansas to the deserts of Utah lifted into a huge dome, rising five thousand feet above sea level.[13] Erosion near the mountains wore away several thousand feet of the sediments deposited in the sinking basins, leaving a broad valley between the High Plains and the Rocky Mountains. Streams along the mountains changed course and began to flow northward and southward, stripping away what was left of the High Plains. This erosion uncovered the ancient surface and created a wide, hilly valley called the Colorado Piedmont.[14]

Erosion continued to wear down the landscape until only scattered ridges and peaks remained. Between thirty-five and five million years ago, a third set of Rocky Mountains was formed by uplift at the faults and accompanied by volcanic activity. It was this uplift that raised the Pikes Peak Region to its present mile-high elevation, with the tallest peaks nearly three miles high.[15] The earth grew colder, ushering in the Pleistocene Ice Age three million years ago. The accompanying glaciation eroded the high country and the Colorado Piedmont.

Grand Sentinel
(Towers and Pinnacles),
Photo, Colorado College

Balanced Rock
Author's Collection, Photo
by Evans

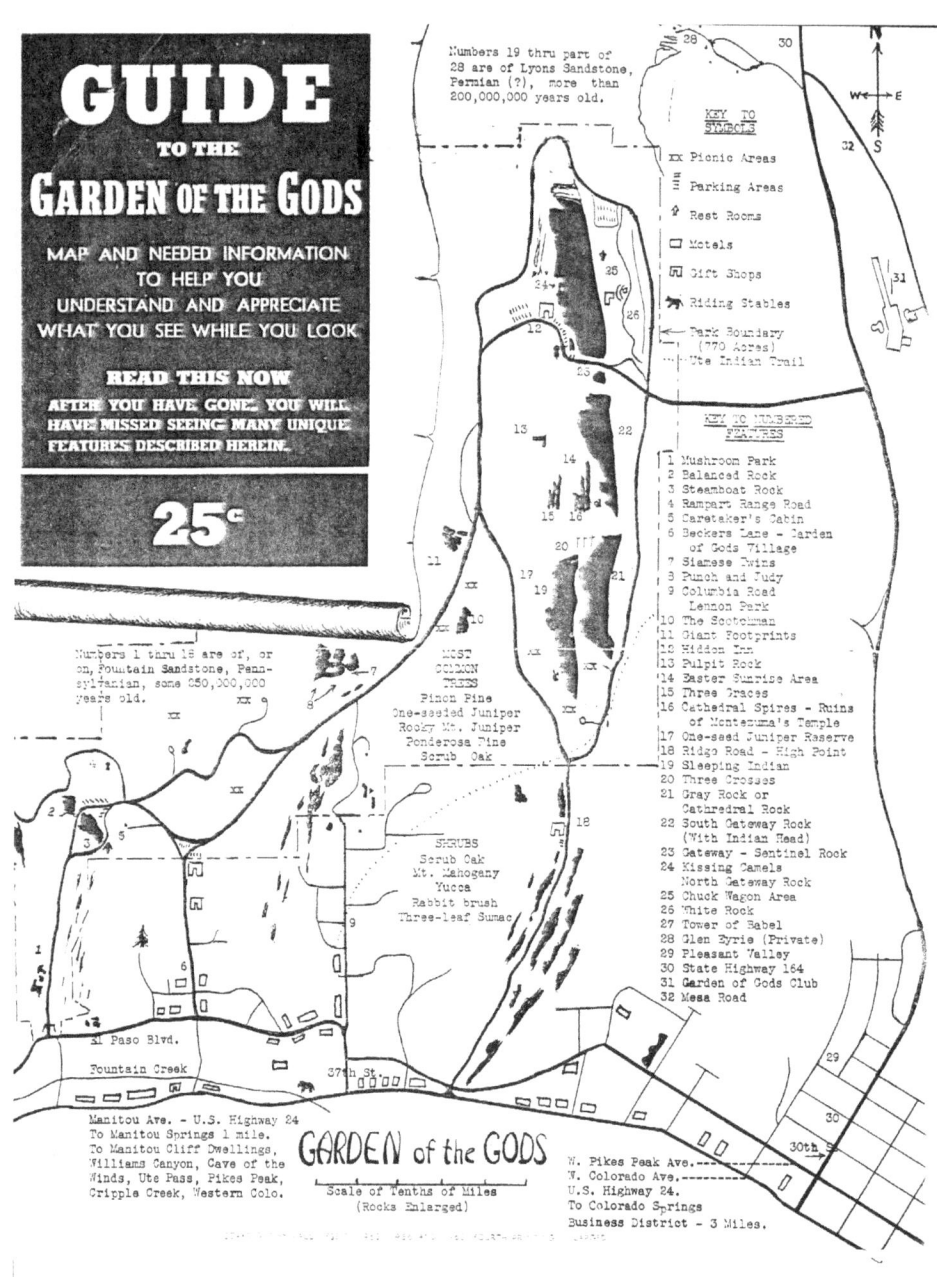

Guide to Garden of the Gods,
Guide, Paul Nesbit

White Rock
Photo, Manitou Springs
Heritage Center, Clinton

This erosion carved the rocks of the Fountain and Lyons formations bent up by the mountain uplift and created the spectacular towers, pinnacles, and mushroom-shaped rocks of the Garden of the Gods.[16] The rocks of the Fountain Formation, named for nearby Fountain Creek, may be seen throughout the Garden of the Gods area. The red color comes from the mineral hematite which, along with silica, bonds the coarse sands and gravels together.[17] The Balanced Rock is a good example of the Fountain Formation.[18]

The lighter colored Lyons Formation is made of fine-grained sand. The Tower of Babel and White Rock which seems to have been formed from the sands of an ancient beach but lacks fossils are from the Lyons Formation. The various rock features in the Garden of the Gods area have five different classifications (slanted, toppled, stood-up, pushed around, and overturned) depending upon which fault or forces of nature acted upon them to determine their positions.

The slanted rocks, varying in the degree of their slants, were bent upward by tremendous forces along the faults and uncovered by erosion, which also gave the rocks their present shapes. These may be seen looking from Sleeping Giant toward Gateway Rocks, and when viewing Giant Footprints.

Slanted Rocks
Photo, Author's Collection,
Evans

The toppled rocks have fallen down from projecting ledges through the action of wind and erosion. Each has settled at its own new tilt, some even upside down. Some of the pedestal rocks such as Balanced Rock and Toad and Toadstools originated this way. Their weight compacted the soil underneath, and their bulk protected them from weathering. When the surrounding soil was washed or blown away, the toppled rock was left on a pedestal.

Mushroom Garden
(Toppled Rocks),
Photo, Manitou Springs Heritage Center

Cathedral Spires (Stood up Rocks),
Photo, Author's Collection, Evans

Pushed-around Rocks,
Photo, Colorado College

Tower of Babel, (Over-turned Rocks)
Photo, Author's Collection, Evans

Fountain qui Boille
Photo, Manitou Springs Heritage Center, Brent

The Gateway Rocks and the Three Graces are examples of stood-up rocks. A great, upward force broke the rocks along the Rampart Fault and turned them to a vertical position.

The pushed-around rocks seen just east of Sleeping Giant became stood-up rocks near the Lyons Fault (parallel to and just west of Sleeping Giant) before they were caught in the movement of that fault and turned until they pointed nearly southwest.

The over-turned rocks seen in the Tower of Babel were turned beyond the vertical position until the older layers were somewhat above younger.[19] The faults that created the unusual geology of the Garden of the Gods area also created mineral springs.

Springs of mineral water bubble to the surface through a number of channels in the shattered rock along Ute Pass Fault. The water from these springs is naturally carbonated, having absorbed carbon dioxide from carbonate rock (Paleozoic limestone) at a depth where pressure is similar to that in a capped soda bottle. As the water rises rapidly along the Ute Pass fault zone, pressure on it decreases and the carbon dioxide comes out of the solution to form the bubbles that give the water its refreshing fizz.[20] The bubbling springs contribute water to Fountain Creek, which was formerly named Fontaine qui Bouille, fountain that boils. The town of Manitou Springs was located here to take advantage of the therapeutic water.

Shoshone and Bottling Plant
Photo, Manitou Springs Heritage Center, Stromer

Ernest Ingersoll, in his book *The Crest of the Continent*, speaks of six springs varying in temperature from 43 degrees to 56 degrees Fahrenheit, "...all strongly charged with carbolic acid." The first spring he writes of is the Shoshone, close to the main road, also called the Sulfur Spring, from the yellow deposit around it. The next spring was a few yards further on and in a ledge of rock overhanging the right bank of the Fountain. This was the Navajo, which contained carbonates of soda, lime, and magnesia, and was more strongly charged with carbonic acid than the Shoshone. The Navajo had a refreshing taste similar to seltzer water. Across the stream from the Navajo was the Manitou. Its taste and properties nearly resembled the Navajo.

Up Ute Pass about one quarter mile on the right bank of the Fountain is the Ute Soda, which resembles the Manitou and Navajo springs but is chemically less powerful and very refreshing. On Ruxton Creek in a tributary valley of Manitou is the Iron Ute Spring. This water is effervescent, cool at 44.3 degrees Fahrenheit, and said to be agreeable despite its strong taste of iron. Further up the left bank of Ruxton Creek is the Little Chief Spring. This spring is less agreeable

Child in Cart (Navajo)
Manitou Springs Heritage Center, Brent

Iron Ute Spring
Manitou Springs
Heritage Center,
Brent

in taste, less effervescent, and more strongly impregnated with sulfate of soda than any of the other springs. It contains nearly as much iron as the Iron Ute Spring. [21]

Today the Manitou Springs Chamber of Commerce and Visitors Bureau reports the location of four more springs; the Twin Spring on the left side of Ruxton Avenue past Osage Avenue, Stratton Spring at the corner of Ruxton Avenue and Manitou Avenue, Wheeler Spring on the left side of Canon Avenue north of Park Avenue, and Seven Minute Spring off El Paso Boulevard north of Mansions Park. These springs with their therapeutic qualities have attracted people and animals for thousands of years.

The Garden of the Gods area with its colorful mix of dramatic rocks as the centerpiece has also attracted people and animals because of its physical location. It is located at an important geographical, biological, and ecological convergence. To the east is the prairie, to the southwest is arid foothill and canyon country, and to the west and northwest are the rugged higher mountains.[22]

The Great Plains and the Rocky Mountains, two of the nation's largest geographic features, meet on a north-south line through Colorado at the Garden of the Gods. Pikes Peak stands farther east and rises more steeply from the Great Plains than does any other 14,000-foot peak. This

Little Chief Iron Spring
Manitou Springs
Heritage Center

brings the Alpine Zones (timberline, persistent snowbanks, and other high-mountain features) within easy reach and close viewing from the Garden of the Gods.[23] This geographical convergence is not the only unusual aspect of the Garden of the Gods area.

At first glance, it seems that little vegetation of interest may be found among the red rocks; however, the Garden of the Gods area is a center of unusual plant and animal associations. While common for two of the three great plant associations or zones to meet for long distances, the Garden of the Gods area is unusual in that all three very distinctly meet here. These zones are the Upper Sonoran (Pinon-Juniper Woodlands), the Transition Zone (Ponderosa Pine Woodlands), and the Brushland (Scrub Oak and Mountain Mahogany). Generally, the vegetation zones change with altitude and latitude. A gain of approximately 2000 feet in altitude equals a gain of about 10 degrees in latitude. From zone to zone, the average temperature drops about 7 degrees.[24]

In addition to being on the line where the Great Plains meet the Rockies, the Garden of the Gods, Glen Eyrie, and Red Rock Canyon are also at the point along that line where the typical southwestern forms of vegetation penetrate farthest north. Trees, or any form of mountain life, may be found living higher or farther north on sunny slopes and lower on shady slopes than would otherwise be normal. The Pinon-Juniper Woodland reaches this far north because of the generally south-facing slope in the Garden of the Gods.

One can follow the Upper Sonoran Pinon-Juniper Woodlands from south-facing slopes two miles north of Queens Canyon at Glen Eyrie southward and westward around

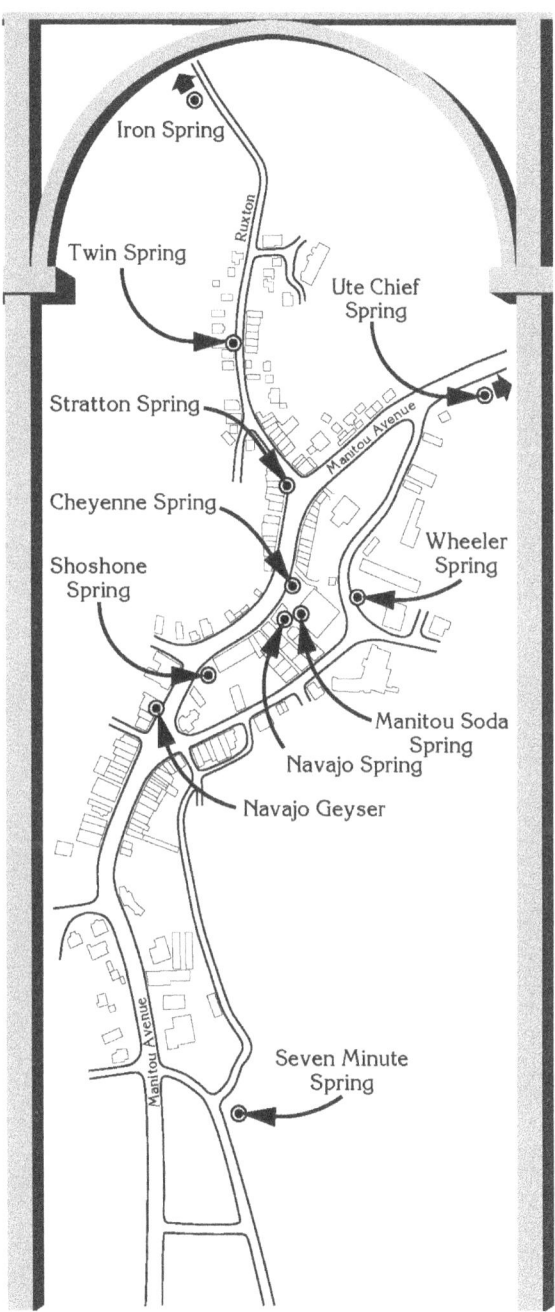

Map of Mineral Springs
2009, Manitou Springs Heritage Center

Gateway (Eco-Zones)
Manitou Springs Heritage Center, Hook

the foot of the mountains into New Mexico, Utah, and Arizona. The transition zone of Ponderosa Pine, Douglas fir, and Colorado blue spruce forests may be seen on the western and northern borders of the Garden of the Gods and Glen Eyrie. One finds the Brushland zone, typified by scrub oak and mountain mahogany in the foothills north and west of the Upper Sonoran zone and on north-facing slopes in the Garden of the Gods below the Transition zone.[25]

Each of these zone associations or eco-niches includes not only plants, but many animal forms dependent upon those plants for food and shelter. The Garden of the Gods proper has only one small spring. However, the adjacent Glen Eyrie with its north-facing slope and stream-bottom makes the habitat more complete and provides a wide variety of plants and animals.[26]

Because this unique area holds such a fascination for modern peoples, it is logical that ancient peoples would have been in awe of its grandeur as well. Primitive peoples must have been

Kissing Camels and Pikes Peak (Eco-Zones),
Colorado College, Beidleman

grateful for the wide variety of vegetable and animal life and the readily available shelter provided by the rocks in the place we now call The Garden of the Gods.

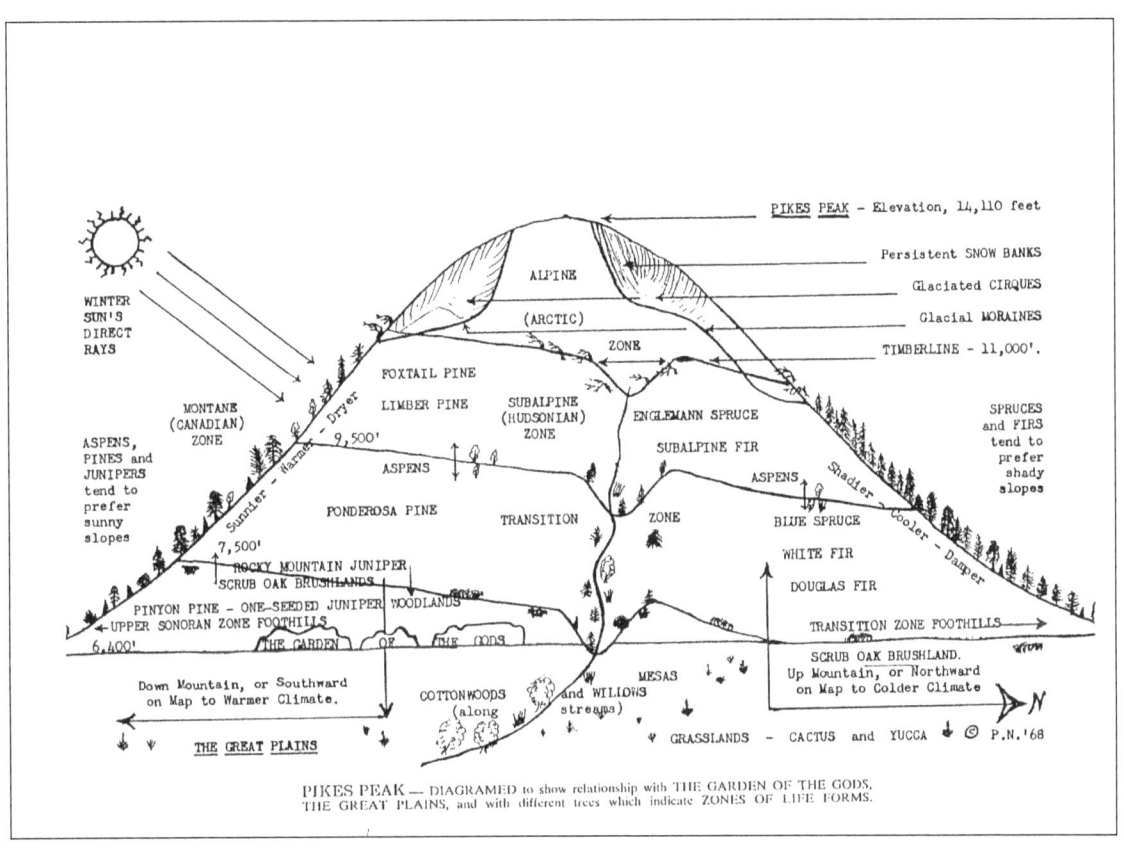

Zones of Life Forms,
Map, Paul Nesbit

Endnotes:

1. Chronic, Halka, *Roadside Geology of Colorado* (Missoula, Montana: Mountain Press Publishing Co., 1980), xiii.
2. *Ibid.*, 4-5.
3. Walker, Melissa. "Clues to Prehistoric Life and Landscapes of Garden of the Gods Park." (Colorado Springs, Colorado) 1998.
4. Chronic, xiii.
5. Walker, Melissa. "Clues".
6. *Ibid.*
7. Col, Jeananda, Zoom Dinosaurs. http://ZoomDinosaurs.com 1996.
8. Walker, Melissa. "Clues".
9. Bill Reed. "Fossil Fuels New Discovery: Scientists name species after Garden of the Gods." *The Gazette.* 25 May, 2008, Life 10.
10. Col, Zoom Dinosaurs.
11. Chronic, 19.
12. *Ibid., 89.*
13. *Ibid.*, 21.
14. *Ibid.*, 21.
15. "How Did Those Red Rocks Get There?"
16. Chronic, 5.
17. *Ibid.*, 91.
18. Nesbit, Paul A., *Garden of God's* (Colorado Springs: Paul A. Nesbit), 10-13.
19. *Ibid.*, 11.
20. Chronic, 90.
21. Ingersoll, Ernest, The Crest of the Continent: A Record of a Summer's Ramble in the Rocky Mountains and Beyond (Chicago: R.R. Donnelley and Sons, Publishers, 1885), 38-40.
22. Nesbit, 37.
23. *Ibid.*, 37.
24. David Carey, "1993 Garden of the Gods Archaeological Survey." Colorado Springs: City of Colorado Springs Parks and Recreation Department, 1993. 2.
25. Nesbit, 19
26. *Ibid.*, 35.

Prehistoric Cultures

How long ago did people use the Garden of the Gods? Archaeological evidence found in the Garden of the Gods in 1993 indicates a presence of people in approximately 1330 B.C., and a radiocarbon dating in 1996 indicates seasonal encampments in the Garden by around 250 B.C. Bifacial stone scrapers, chert (flint) flakes, fire-cracked rocks, possible grinding stones, and a small shard of pottery were found while excavating several hearths and a rock shelter in July 1996. The hearths were not those of a living area because no charred animal bones were found nor was there any other evidence of long-term occupation. They were more than likely used in spiritual rituals. Who made these hearths? Judging from the evidence, the peoples using these encampments were of the middle- to late-archaic period (3,000 B.C. - 500 A.D.) and of the early Ceramic/Woodland period (1 A.D. - 1,000 A.D.). The physical environment at that time was similar to what is found in Colorado today.[1]

The archaic peoples were hunters and gatherers who camped under rock overhangs which are plentiful in the Garden. They hunted small game (deer, antelope, rabbit) as well as big-game animals, foraged for vegetables and grain, and developed a wide assortment of tools.[2] These people used the atlatl (a tool designed to enhance spear throwing), mulling stones, basketry, rabbit nets, stone drills, scrapers, and bone awls. They heated stones in their hearth fires and then used them for heating their food in skin or fiber bags and possibly as a source of heat for small shelters.[3]

These archaic hunter-gatherers had to collect enough food for immediate needs and for storage for winter. Food spoilage was a problem due to bacteria and rodents. Traveling on foot limited the quantities and types of food that could be carried, and winters were deadly if stored food did not last.[4] Many of these people moved with the seasons in order to follow game. Over time they developed semi-permanent settlements of less than twenty-five people and evolved more complicated social patterns. The primary factors determining settlement locations included availability of water, good vantage points, and rich floral and

People of the Garden Time Line
8000BC - 1997AD

	6000 B|C	3000 B|C	1 A|D	500 A|D	1000 A|D	1500 A|D	1600 A|D	1700 A|D	1800 A|D	1900 A|D	2000 A|D
Vail Pass Burial		x									
Middle-Late Archaic				xxxxxxxxx							
Woodlands to Plains					x						
Ceramic/Woodland					xxxxxxxxx						
Fremont People						xxxxxxxxx					
Burial (discovered 1994)						xxx					
Proto Historic							xxxxxx				
Utes							xxxxxxxxxxxxxx				
Spanish Explorers								xxxxxxxxxx			
Apache								xxx			
French Explorers								xxxxxxx			
Comanche									xxxx		
American Explorers									xxx		
Arapahoe									xxx		
Cheyenne									xxx		
Miners, Settlers, & Entrepreneurs											xxx

People of the Garden Time Line, Paul Nesbit

faunal resources, all of which are available in and around the Garden of the Gods.⁵ Gradually, the Archaic people engaged in rudimentary agriculture, traded with their neighboring communities, and made crude pottery.⁶ It is uncertain whether these people were of a distinct mountain culture or if they were of a plains culture and came into the foothills and mountains in search of game, to gather plants, or for ceremonial purposes.

The Ceramic or Woodland culture, characterized by cord-marked pottery and small, side- and corner-notched points as well as other stone tools,⁷ appeared in Colorado around 100 A.D. These people, like the Archaic people, appear to have subsisted on bison, elk, deer, antelope, rabbits, and other small game, and an assortment of wild plants. They collected wild seeds such as goosefoot and pigweed for additional sustenance. The Colorado Woodland people may have begun cultivating corn toward the end of the period; however, archaeological evidence indicates they were apparently not enthusiastic horticulturalists. Hunting and gathering were still their primary sources of food. Men would hunt with atlatls or bow and arrow. They occasionally hunted in large, cooperative groups, driving the game into places where it became immobilized and easy to kill.⁸ Archaeologists have found evidence in the Garden of the Gods indicating the butchering of bison and of medium-sized mammals, probably deer or bighorn sheep.⁹

The transformation from the Late Archaic to the early Woodland Period was simply a matter of adding the new

Ancient Rock Shelters
GoG 1996 Field School, University of Colorado Colorado Springs, Permission of Department of Anthropology

Big Horn Sheep
Author's Collection

skill of pottery making; otherwise, the people retained their hunting and gathering economy. Other Plains Woodland sites are located in the low foothills of the Front Range of Colorado and in the Aurora and Parker areas east of Denver.

Prehistoric people frequented the Garden of the Gods because of its abundance of plant and animal life which would sustain a small population. A survey taken in 1937 indicated the flora included evergreen trees such as white fir, red cedar, yellow pine, Douglas fir and pinon pine, and a few spruce which seemed to have been introduced. Some of the original junipers were estimated to be 700 to 1,200 years old. Deciduous trees and shrubs included Narrow-leaf Cottonwood, Western Broad-leaf Cottonwood, Gunnison Scrub Oak, and Almond or Peach-leaf Willow. Woodbine, Mountain Mahogany, Common Chokecherry, Sumac, Golden Currant, Common Wild Rose, Snowberry, Low Ninebark, Small Fruited Gooseberry, and Yucca plants were found in profusion. Wildflowers included Anemone, Columbine, Mariposa Lily, Indian Paint Brush, Scarlet Bugle, Blue Bells of Scotland, clover, thistle, poppy, Common Milkweed, Wild Geranium, Golden Aster, cactus, and Townsend Daisy. Grasses included wild western wheat, gamma grass, sweet clover, and a little buffalo grass.[10]

Pinon Pine,
Photo, Author's Collection, Evans

In 1937 a herd of twenty-one deer made their home year-round in the park. Jack rabbits and cottontails, skunks, badgers, chipmunks, prairie dogs, several colors of squirrels, coyotes, rock lizards, and numbers of birds, including swifts who made their home in the upper reaches of the gateway rocks were also sighted.[11] The Garden of the Gods is also home for colonies of rare honey ants, one of the world's most unusual species of ants because of its ability to store its food supply (nectar) in its abdomen.[12] This abundance of vegetation and wildlife has attracted people to the Garden of the Gods throughout history.

IMPORTANT GEOLOGY EVENTS

M.Y.B.P*	TIME	
Pliocene	2-12	Mountain streams rejuvenated and begin canyon cutting; the Grand Tetons rise along a fault line.
Miocene	12-25	Extensive erosion in Rockies, intermontane basins fill
Oligocene	25-38	The entire Rockies are above sea level, ending the Laramide Revolution
Eocene	38-55	Last major mountain building pulse in Rockies - Uinta Mountains and most Wyoming ranges form.
Paleocene	60-70	Front Range is uplifted and intermontane parks form. Garden of the Gods is tilted up.
Cretaceous	70-135	Extinction of dinosaurs. A second major mountain building pulse begins in Rockies, beginning the Laramide Revolution. Purgatoire, Dakota, Benton, and, Niobrara Formations are deposited.
Jurassic	150	The Morrison Formation is deposited in tropical lowland.
Triassic	180-225	Ancestral Rockies are completely eroded and buried under their own debris.
Persian	225-270	Lyons Sandstone and Lykins Formations are deposited.
Pennsylvanian	275-310	A major period of mountain building beginning as the Ancestral Rockies develop. Fountain Formation is deposited as aluvial fans along flanks of mountains.
Mississippian	305-350	There is restlessness and early mountain uplift in Rocky Mountain trough.
Precambrian	600	Rocky Mountain trough is initiated.
	1000	Pikes Peak Granite is formed.

*MILLIONS OF YEARS BEFORE PRESENT

MAMMAL LIST

as compiled by Department of Biology, Colorado Gllege, Colorado Springs, Colorado

CARNIVORES:

* American badger - *Taxidea taxus* Raccoon - *Procyon lotor*
* Red fox - *Vulpes fulva* Grey fox - *Urocyon cinereoargenteus* Bobcat - *Lynx rufus*
* Mountain lion - *Felis concolor* Coyote - *Canis latrans*
* Blackbear - *Ursus americanus*
Striped skunk - *Mephitis mephitis*

HOOFED ANIMALS:

Mountain sheep - *Ovis canadensis*
Mule deer - *Odocoileus hemionus*

RABBITS:

Black-tailed jack rabbit - *Lepus califomicus*
Nuttall's cottontail - *Sylvilagus nuttallii*

RODENTS:

Rock Squirrel - *Spermophilus variegatus*
Golden-mantled ground squirrel - *Spermophilus lateralis*
Thirteen-lined ground squirrel - *Spermophilus tridecemlineatus*
Pinon Mouse - *Peromyscus truei*
Deer mouse - *Peromyscus maniculatus*
*Rock mouse - *Peromyscus difficilis*
Western harvest mouse - *Reithrodontomys megalotis* Wood rat - *Neotoma sp.*
Colorado chipmunk - *Eutamias quadrivittatus* Least chipmunk - *Eutamias minimus*
*Pine squirrel - *Tamiasciurus hudsonicus* Porcupine - *Erethizon dorsatum*

Note: This list includes only species for which there has been published documentation. Starred species are uncommon. There should be a number of additions, for example, species of bats, if additional mammalogic field work were undertaken.

1994 Inventory of Wildlife, Important Geology Events, Garden of the Gods Master Plan

Top: Points UCCS/CHS
Bottom: Cord-Marked Pottery, UCCS/CHS

Top Right: Petrified Wood Scraper, UCCS/CHS
Middle: Bone Tool, UCCS/CHS
Bottom Right: Bone Awls, UCCS/CHS

In August of 1994, a hiker found a prehistoric burial site in close proximity to the Garden of the Gods. The remains were consistent with those of a Native American male of less than thirty years of age at death, of seemingly good health with no clear indication of cause of death. The excavation recovered cultural material consisting of stone and bone tools and ceramic shards. The individual was buried in a flexed position, and several of his bones and some tools were stained with red ochre. An examination of one of his tools seemed to indicate it was used to make pottery. A radiocarbon assay on charcoal taken from the burial pit indicated the burial was approximately 1300 years old, placing this man in the Garden of the Gods area between A.D. 680 and 950. The burial site also yielded a later occupation radiocarbon dated from A.D. 1015 to 1265.[13] The Ute claimed this burial, but since the individual was buried in a flexed position and the Ute did not usually bury their people this way, some doubt still exists as to the ancient Indian's origin.

From 1000 A.D. to around 1300 A.D. the population in the foothills of the Colorado Rockies decreased, but after 1300, the occupation increased again. Eastern Colorado, which includes The Garden of the Gods, was inhabited by groups of hunters and gatherers or part-time horticulturalists who acquired the horse and ranged farther afield for subsistence.[14] Pottery is evident, and points have been found which indicate the introduction of the bow and arrow. Other cultural evidence includes debris from the crafting of stone tools, stone circles or tipi rings, and hearth features. However, because of the lack of absolute chronological dates, archaeologists find it difficult to determine which specific group of people left this evidence.[15]

Through their oral traditions, Ute Indians claim a connection to the Garden of the Gods from their creation. However, linguistic and archaeological evidence as stated in the Numic Fan Theory (migration patterns based on the spread of languages) suggest the Ute migrated from their southwestern Great Basin homeland, which includes present day eastern California, Nevada, much of Utah, and parts of Oregon, Idaho, and Wyoming.[16] This hypothesis places the Ute people in the Colorado area no earlier than 1608. Archaeologists, anthropologists, and other scholars have determined that prior to 1600, and for more than 11,000 years, the historic area of occupation by the Ute was the home of several different groups of prehistoric peoples.

According to Cassells in *The Archaeology of Colorado*, there is a strong possibility of migrations into Colorado from several directions. People and ideas entering Colorado after about 1000 A.D. seem to have come by way of the South Platte river system in the north and the Arkansas river system in the south.[17] Ethnographic accounts (anthropology dealing descriptively with specific cultures) indicate occupations along the Front Range of Colorado by groups such as the Ute, Shoshone, Arapaho, Comanche, Cheyenne, Apache, and Pawnee.[18]

Even though lacking archaeological evidence, the Ute Indians have oral traditions of occupying the Colorado area from prehistoric times. According to Alden Naranjo, Southern Ute Tribal Historian, they have stories about the flood, of how the magpie and the different birds or animals came to be, and how man, *Nootch* (the Ute), helped them. They have stories about *Be-do-wah*, great snakes or lizards, and *Be-aduh,* a big fish. Alden Naranjo also tells of hearing his grandfather talk about a bear who is the grandfather of bears. When Naranjo asked his grandfather why he was talking about this bear, his grandfather replied that he's bigger and he's white and the leader of the bears. The grandfathers also talked about hunting the *Ah-ree-put*, an elephant with a long nose and big tusks. They also tell of European visitors prior to the arrival of Spanish explorers.

In 1989 in the White River National Forest of Colorado, a team of archaeologists excavated a cave burial. This cave was at 10,500 feet and the radiocarbon analysis put the age of the burial at 8,000 years old, a time when the Rocky Mountain valleys were still boggy with melting glaciers. The excavating team believes the man and his small band of about thirty people lived in that rugged environment year-round. According to Alden Naranjo, the Ute creation story tells us the "Utes were created and that they would always live in the mountains, up high in the mountains, that they would be closer to *Se-no-ee,* the creator. Some would go down into the valleys and be there, but they were always going to be up in the mountains. The Utes were agreeable to the creator because they did not follow the crazy coyote, the creator's little brother, so they would always be in the mountains close to the creator."[19]

Naranjo remembers stories which were told many years ago of how the Ute people used to travel from the front range, down towards the Spanish Peaks, over the mountains, and then down into what is now New Mexico, towards a pueblo called Albique, and back around. They visited the pueblos through that part of the country and made contact with some of the pueblos that live in the cliffs, *Moo-ke-chee,* with whom they were friendly. The Ute would also meet up with their allies, the Jicarilla Apache, and go into Navajo country and come back home the same way.[20] Archaeological evidence in the form of pottery indicates the Anasazis also traveled up to Colorado for purposes of trade.

Naranjo tells another story which is related to the Ute visits to the cliff dwellers, and is more specific:

There was this old lady that was the last of her people from... it was from Mesa Verde or Hovenweep, one of those places, it might have been Mesa Verde. She was the last of her people. She was captured when she was a little girl, she was captured in that area so she was married to somebody from the group. The story goes that the evenings when she got old, when she got real up in age, maybe 80 or 90 years old, she used to sit up on this hill and face towards her people and talk to them in that Moo-ke-chee language, that she remembered and she was an old lady when she passed away but she was the last of her people. She knew that her people had moved on, you know, when she was the last of her people that still remembered her traditional ways, her people's ways. Maybe she was around 12 or 14 years when she was captured... been a Ute for the rest of her life. She had become married and also had children from the man that she married and she was torn between going back to her people or staying with her people because she had that freedom after she got a certain age and became accepted. But she never did because she had children, grandchildren. So she didn't know her people anymore and she didn't know how they would accept her. She stayed, but she used to do that, sit on the hill... cry for her people cause her people she had were gone. They had moved on or whatever, she was the last of them. She was the last one.[21]

This would place the Ute in this area around 1200 A.D. or a little later if this woman was the last of her people, the Anasazis.

Stories or oral traditions about possible Norsemen, bearded, red headed, blond people coming to this part of the country, way before the Spanish ever came are also a Ute tradition. The stories tell about how these red headed and blond people walked to this country. They say that these people used to have *Wa-nesh-gee-ga,* iron-headed horns, or horns on their metal headpieces. Some of the old petroglyphs and pictographs depict these people.[22] These Ute oral traditions are clearly speaking of times prior to the first Spanish contact in the sixteenth century and appear to support a hypothesis presented by Dr. Goss in 1977.

Dr. Jim Goss, a noted anthropologist from Texas Tech University who has made extensive study of the Ute culture, wrote a paper in 1977 hypothesizing that the ancestors of the Ute have been in the western mountains for the last 10,000 years. He explains that the heartland of the ancestral group of Numic-speaking people extends from the Sierra Nevadas to the Front Range of the Colorado Rockies. Over the centuries, they have taken refuge in their mountains during times of stress, and then moved back out onto the prairies and desert lands when conditions improved.[23]

Rock art found throughout Utah and Colorado is the evidence supporting prehistoric occupation by the Ute people. According to the archaeologist William Buckles of the University of Southern Colorado, "Some rock art styles clearly identifiable as historic Ute manufacture differ little from earlier prehistoric styles of art." According to Cassells, Ute rock art differs from Anasazi and Fremont rock art (also found in Colorado), and can be found on panels across western Colorado.[24] George Beckwith was told in 1931 by Ute in Sevier County, Utah, that rock writing was *wee-noose-a-pope*, which means writing by ancient Indians or old-time Indians.[25] Bill Buckles draws the conclusion from his Ute Prehistory Project that the later Desert Culture, which is analogous to the late Archaic Culture, is ancestral Ute, since little difference can be detected in the lifestyles of the two groups.[26] Throughout the southwestern United States are examples of rock art, some dating at least to the Archaic period.[27]

For centuries, people who have visited the Garden have felt the need to place their marks on the soft sandstone rocks. Ancient rock art in the form of petroglyphs has been found in the Garden of the Gods. According to Alden Naranjo, all the tribes that have been in the area of the Garden of the Gods have their own stories about the petroglyphs, including the Ute. This particular rock art panel consists of five solid-pecked petroglyphs, consistent with the Umcompahgre style of rock art. Scenes of what appear to be men with raised arms standing before and near quadrupeds such as deer, elk, bear... are common at sites throughout west-central Colorado and may be symbolic of shamanism and hunting rituals. Such a scene is depicted in the petroglyphs in the Garden of the Gods.[28]

As time passes, scientists are discovering more and more evidence that seems to correspond with the oral traditions of the Ute people. Perhaps, as they say, they were always here and have simply evolved on their own or learned new skills through trade. Perhaps some day archaeologists will discover cultural artifacts that substantiate the Ute Indians' claims that their ancestors are the only peoples to have occupied these lands of Colorado until other historical Indian tribes migrated into their territory.

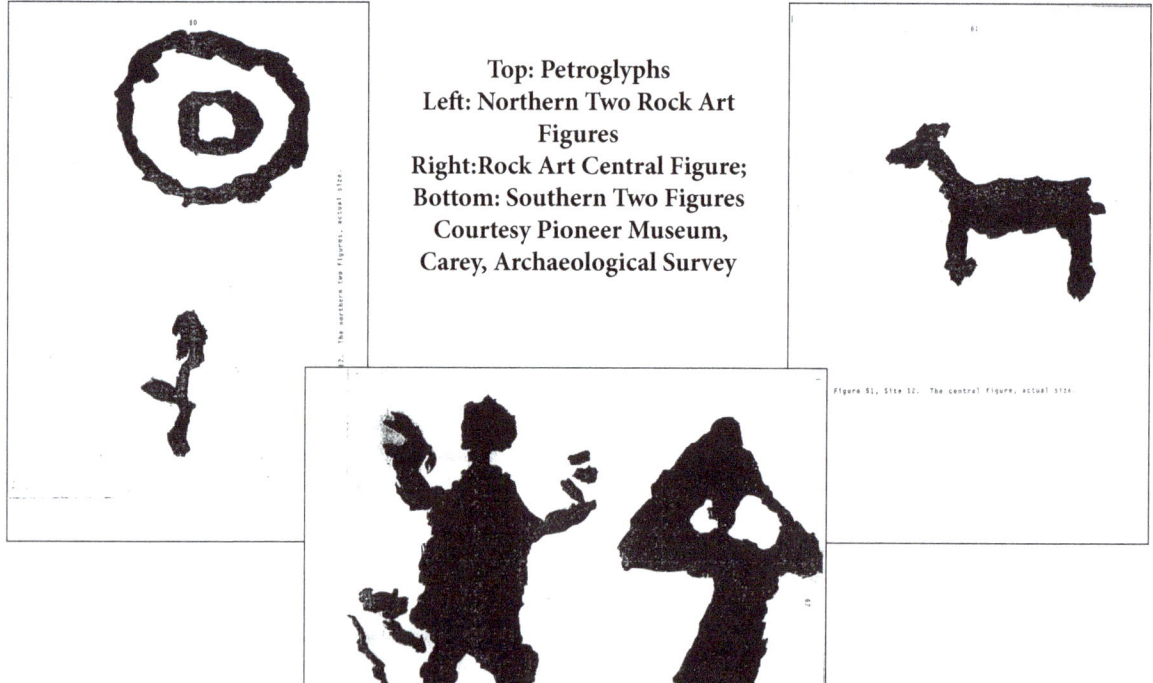

Top: Petroglyphs
Left: Northern Two Rock Art Figures
Right: Rock Art Central Figure;
Bottom: Southern Two Figures
Courtesy Pioneer Museum, Carey, Archaeological Survey

Endnotes:

1. E. Steve Cassells, T*he Archaeology of Colorado (*Boulder: Johnson Books), 101.
2. James A. Maxwell, ed., *America's Fascinating Indian Heritage*, (Pleasantville, N.Y.: Reader's Digest Association, Inc., 1978), 19.
3. Carey, 8.
4. Jan Pettit, *Utes The Mountain People* (Revised Edition) (Boulder: Johnson Books, 1990),7-8.
5. Guthrie, et al., *Colorado Mountains Prehistoric Context* (Denver: State Historical Society of Colorado, 1984), 34-35.
6. Maxwell, 24.
7. Jeffrey L Eighmy, Colorado Plains Prehistoric Context (Denver: The State Historical Society of Colorado), 83.
8. Cassells, 163-164.
9. Carey, 9, 13.
10. Bruce A. Woodard, *The Garden of the Gods Story* (Colorado Springs: Democrat Publishing Co., 1955), 12.
11. *Ibid.*,
12. "Rare Honey." *The Colorado Free Press*, 20 April 1962, as reported in The Colorado Prospector: Colorado History From Early Day Newspapers, Vol. 17, No 4.
13. William R. Arbogast, Forest D. Tierson, Alden Naranjo, *A Prehistoric Burial at 5EP2200*, El Paso County, Colorado (Colorado Springs, University of Colorado the Springs, 1996), ii.
14. Eighmy, 21-22
15. Guthrie, et al., 46.
16. Maxwell, 25.
17. *Ibid.*, 180-181.
18. Guthrie, et al., 48.
19. *Ibid.*, 17.
20. Smith, Eugene R. Mystery Man (Denver: The Colorado Historical Society, 1996), 16.
21. *Ibid.*, 16.
22. *Ibid.*, 18
23. *Ibid.*, 24.
24. Cassells, 191-192.
25. Pettit, 9.
26. Cassells, 191.
27. Eighmy, 75.
28. Carey, 25.

American Indians and the Garden of the Gods

Because American Indian history is an oral tradition, it is difficult to obtain written historical documentation concerning their relationship to the Garden of the Gods. The Indian people are generally reluctant to share their stories with outsiders, perhaps from fear of exploitation. A long history of broken promises that continues even today makes it difficult for the American Indian people to trust white men. As recent as 1994 at the Garden of the Gods American Indian Workshop, a promise was made to the Indians in attendance that a large area in the new Garden of the Gods' visitors' center would be set aside for the purpose of telling the history and culture of the Indians who were in the Garden of the Gods and the Pikes Peak area. To this date, this has not happened. Annette Frost, Southern Ute elder, said that while she was growing up, her grandparents told her:

Don't you tell anything to anybody. The White man. Don't tell him stories about anything... I have grown up, and I'm about 80 years old today, and I have kept a lot of these things in the back of my mind. I have never brought it out. There are just a few things that I have brought out for my children. For them to hear these things. I have told my daughter many things. And, I have also told her that there are a lot of ways that you cannot go out into the whole thing, and tell the White man about. Every Indian that I know of has told me these things. Don't go out and tell everything to the White man. They will write all these things and put it in their own version. Things that we shouldn't even say.[1]

Alden Naranjo, tribal historian of the Southern Ute, said, "If you say it [the stories] in our language, sometimes we have to interpret it into the White man's language and it comes out differently. But how we interpret it is what comes from our hearts. Of how we respect the ground that we walk on. How we respect the things that are here... There are places throughout the country where Indians have placed these things [mounds and

Medicine Wheels] for a specific reason. There are places like Devil's Tower, Bear Butte, Garden of the Gods. Those are places that are special places for the Indian people."[2] The Ute are the most willing to share some of their stories about the Garden of the Gods with outsiders, which perhaps is appropriate because they have the greatest historical connection with this area.

At least eight American Indian Nations have an historical relationship with the Garden of the Gods. The Ute, Apache, Kiowa, Shoshone, Comanche, Cheyenne, Arapaho, Pawnee, and Lakota Nations all claim a connection to the Garden of the Gods. What drew these people to the Garden of the Gods and what was the nature of their relationship with it and with each other?

The American Indian concept of land ownership was quite different from the European concept. No one owned the territorial range the Indians covered, nor did they own the natural resources of the area. Although the idea of possession in relation to land and water was alien, the various Indian groups or bands did have a concept of territory. The rights to use and manage a given territory were vested primarily in the band or village population who customarily lived, hunted, gathered, and traveled in that area.[3]

These territorial rights overlapped, and many were shared by members of different groups. Local bands that customarily occupied and traveled a given area became identified with that region and its use. However, this persistent use did not exclude outsiders from coming into the region to use it temporarily or even on a more permanent basis.

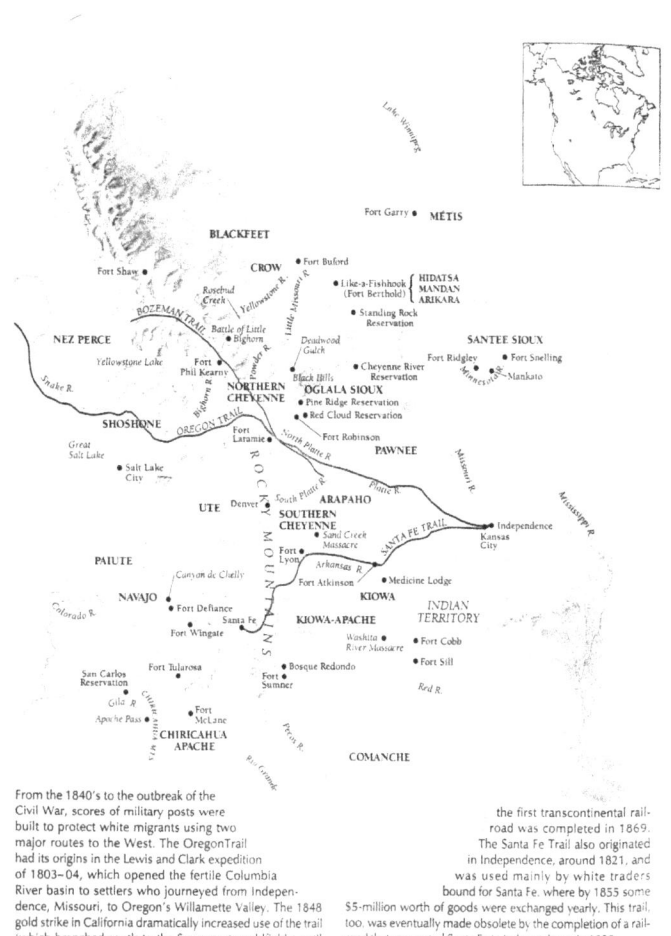

Indian Territory, Cassidy et al. *Through Indian Eyes*. Pleasantville New York: Reader's Digest Association, Inc. 1995. 300

Long-term patterns of inter-group sharing of lands and resources appear to have been widespread. Some were of short duration, others lasted over a century.[4]

When groups were military allies and related to each other through extensive kinship ties, they shared access to the use of a commonly held territorial range. The territorial spacing of tribes was a reflection of whether a tribe was at war or peace with its neighbor. Tribal boundaries were not necessary between allied tribes, and it was common for an area of neutral land to separate hostile tribes, lands through which they passed en route to make raids on enemy camps.[5]

Several Indian tribes claim a connection to the Garden of the Gods, Glen Eyrie, Red Rock Canyon, and the boiling springs at Manitou. However, the Ute, who have a longer continuous history in Colorado than any other tribe, have the strongest connection. Their traditional lands occupied all of the mountainous lands in Colorado from the Yampa River in the northwest to the San Juan River in the southwest, and from the slopes of the Front Range in the east to the present western border. The Ute hunted buffalo as far out onto the plains as the shadows of the Rockies stretch at sunset.[6] The Ute homelands also included central and eastern Utah and parts of New Mexico. Today the Ute are the only tribe with reservations in Colorado.

For centuries, the Ute were the only Indians along the Front Range of the Colorado Rockies until the Arapaho and Cheyenne moved into eastern Colorado from South Dakota, pushing the Comanche and Kiowa ahead of them. With the discovery of gold in Colorado in 1850, Euro-Americans came onto the plains and killed off the buffalo, driving the Arapaho and Cheyenne into the mountains and into conflict with the Ute.[7]

The Ute have legends about the Garden of the Gods and the surrounding area. One such legend speaks of the Garden of the Gods, Glen Eyrie, and Blair Athol as being abodes for the spirits:

The Three Spirits
A long time ago when all of the people of the earth had been living in a golden age, their number had multiplied so greatly that they were forced to spread themselves over all the country round about. To help them, there came to live in their midst three of the Lesser Spirits. The first was to teach and help them with agriculture, the second was to help them make weapons, set traps and hunt, while the third was to teach them religion, worship of the one Manitou, and how to govern themselves properly. Each of the Spirits was said to have built for himself a great and magnificent temple for his abode. The identity of the particular temple belonging to each of the three Spirits has been lost down through the years but they are known now by the names of The Garden of the Gods, Glen Eyrie and Blair Athol.[8]

Another Ute legend tells of a great flood that covered the top of Manitou's Mountain; when waters subsided, the floating animal carcasses turned to sandstone and rolled down the peak into the valley below.[9]

Who were the Ute People? They called themselves Nuche meaning the people or we the people. The Spaniards called them *Yutas*, the Cheyenne called them Black People, and they were known as the Rabbit Skin Robes and Deer Hunting Mountain People by the Omaha and Ponca. They were respected and feared by the surrounding tribes.

Unlike other tribes, the Ute have no migration legend. They claim occupation of the Colorado-Utah area from their origin. Their creation legend has several versions, almost all having animals as principal characters. They believe animals used to talk and act like humans. The Coyote appears frequently in the stories, sometimes as a trickster, dupe, or troublemaker.

Ute Origin Myth
Once there were no people in any part of the world. The Creator began to cut sticks and place them in a large bag. This went on for some time until, finally, Coyote's curiosity could stand the suspense no longer. One day while the Creator was away, Coyote opened the bag. Many people came out, all of them speaking different languages and scattering in every direction. When the Creator returned, there were but a few people left. He was very angry with Coyote, for he had planned to distribute the people equally in the land. The result of the unequal distribution caused by Coyote would be war between the different peoples, each trying to gain land from his neighbor. Of all the people remaining in the bag, the Creator said, "This small tribe shall be Ute, but they will be very brave and able to defeat the rest.[10]

Anthropologists believe the Ute are probably descendants of the Desert Culture (8,000 B.C.-A.D. 400), the Fremont Culture (A.D.400-A.D.1200), and perhaps the Basketmakers (A.D. 450-700). If so, their residence dates back 10,000 years or more.[11] The Ute speak one of the Shoshonean (Uto-Aztecan) languages similar to the Shoshones, Comanches, Bannocks, and Paiutes. These tribes are all related to the Great Basin (southern California, southern Nevada, and Utah) tradition. The Ute recognize these other tribes as relatives to a certain extent. More distantly related languages are spoken by the Hopis, Chemehuevi, and some of the California tribes. Nahuatl, the language of the Aztecs of Mexico, is also related to the Shoshonean language.[12]

According to Alden Naranjo, the Garden of the Gods was a traditional winter campground and sometime summer campground for the Mouache band of Ute. The camp consisted of members of other bands of Ute also, the Tabewache and Capota. These three bands were the primary bands in the area of Colorado Springs; however, only the Mouache bands actually lived in the Front Range of the Rocky Mountains, ranging from around Cimarron, New Mexico to the Denver area and east to about the Colorado/Kansas border. This was during the time of the Spanish contact and before the coming of any European, or the migration or movement of any other Indians into the plains. Plains culture had not yet started. After other Indians moved into the plains, the area of Colorado Springs then was shared by the Ute and other Indians.[13]

From early spring until late fall, the bands of Ute would break up into small family units. Each family unit had to have a great deal of room since food was scarce and gathering it could not be done very well in large groups. Late in the fall, the family units would begin to move out of the mountains and into sheltered areas for the winter months. Generally, the family units of a particular band lived close together during the winter. The Garden of the Gods area was one of the winter homes of the Ute.[14]

The Ute generally made camp on a hill or high spot within walking distance of a spring or stream. Homes were established in a grove of trees that furnished protection from storms and provided firewood. Before the Ute obtained horses from the Spanish, they lived in wickiups, domed willow huts about fifteen feet in diameter and eight feet high, covered with willows, juniper bark, and grass or tule. Leather-covered teepees were not practical until after the introduction of the horse because of the difficulty of moving them. The wickiup had a small fire pit in the center for warmth; however, the women did most of the cooking outside when weather permitted. An average of about five persons lived in each wickiup and as many as fifty-five persons lived in one village.[15]

Yucca,
Author's Collection, Evans

The wildlife and abundant plant life in and around the Garden of the Gods made it a desirable place for the Ute to live. Yucca was one of the most useful plants. The blossoms and seeds are edible. The people made soap from the roots, and rope and twine from the fibers in the leaves. The leaves of many plants were used as greens and boiled with meat. The Ute also harvested roots and tubers of plants such as yampa, cama, wild carrot, and wild onion with digging sticks three to four feet long. The women baked the roots by placing them in earthen ovens. The people collected sweet sap from pine trees by slashing the tree and inserting a sharpened hollow deer bone in between the bark and the center. The sap flowed into a bark container and was eaten immediately. When food was scarce, the Ute stripped bark from trees and peeled off the substance inside the bark to eat raw or mix into mush with other ground foods. The pinon tree was important for the nuts which the people ate raw or parched in hot coals to remove the shells or ground into meal and stored for later use. They mixed this meal with water and baked it in cake form in ashes or hot rocks. These nut cakes could be stored for a long time.[16]

Hunting Bison,
Pikes Peak Library District, Poley

After the introduction of horses in the late 1600s, the Ute began to hunt buffalo on the plains. They had the meat for food and the hides for tepee covers, blankets, clothing, moccasins, and bags of all kinds. With the horse the Ute could more easily transport their goods.[17]

Women worked setting up and moving camp; the men got out of the way until the work was done. When breaking camp, the women would send the young boys to get the horses, then the women would take down the tepees and pack up the goods. A witness to a camp move in 1873 described the procedure: "The Utes were moving camp with 400 ponies, many superb animals. Tent poles, six on either side, were fastened to the ponies of the squaws, one end of each pole dragging on the ground behind. The squaws attended to loading and packing the animals. On the top of many of these packs were perched papooses, strapped securely on, but old enough to drive and guide their ponies."[18]

Chase Mellen, brother-in-law of General Palmer who founded

Colorado Springs, recalled the Indians leaving their annual summer camp near General Palmer's home at Glen Eyrie in the late 1800's. He wrote: "I can vouch for the accuracy of that picture of Indian ponies loaded with the squaws and camp equipment, flanked by lodge poles trailing on the ground with baskets strapped securely between, holding the papooses. I can vouch for the truth of the story that when moving camp for a serious purpose the old and decrepit were left behind to die."[19]

Moving Camp
Denver Public Library, Harper's Weekly

The Ute made their camps along Camp Creek outside the Garden of the Gods because the area in among the red rocks is a place of spiritual importance for the Ute people. The spiritual significance of the Garden of the Gods is evident from a story told by Alden Naranjo about the Istubabi, the little people or spirits who lived in the holes in the rocks. "These little people talked to the Ute people, so the Utes did not camp in the Garden of the Gods, they camped along Camp Creek down by the Rockledge Ranch [Whitehouse Ranch], through that area. They also camped back toward Manitou Springs. They did not camp in the Garden of the Gods because it belonged to the little people, the spirits. The Utes believed in those little people and it was a good feeling for them, so in respect for that area, they never camped in there. They went into that area to do whatever ceremonies they had at that time. These ceremonies may not be remembered today; they may not remember how to do them exactly as they used to do them during those times."[20]

Holes in Rocks
Author's Collection, Evans

According to Bertha Grove, a Southern Ute Elder, it is the place where their Bear Dance originated. The Bear Dance is the oldest ceremony the Ute people have. It is held when the bears wake up and it is time to visit relatives, play games, to mourn those who are gone, and to welcome new members of the tribe. During the last day of the Bear Dance, couples are paired off to dance together. Suddenly, the dancing ceases and dancers dressed as a male and female bear spring into the circle, pawing and prancing to symbolize that the prayers have been heard and the earth is awakened.[21]

Bear Dance
Pikes Peak Library District, Poley

During the wintertime, the Garden of the Gods didn't have no snow, when there was snow all around. And during the night, those rocks would keep light up there. And that is why it is very sacred. They had all kinds of dances up in there. And during the night, there were lights up in those rocks... Our parents used to tell us that, about the shining, the light in the middle of the night. We wanted to see it for ourselves, and we did. And we would always pray to it. Our maker gave that to us. This was our place... It used to make us get better. It is very sacred to come over there, and get better there. Young kids that were out somewhere else and got sick, they were brought over there. It made them happy and everything. Annette Frost, Southern Ute Elder[22]

Archaeologists have discovered hearths (fire pits) near the tops of some of the taller rocks in the Garden of the Gods. Perhaps these hold a spiritual significance and possibly were the source of the light in the middle of the night.

The other tribes to frequent the Garden of the Gods and the Boiling Springs at Manitou historically were Plains Indians. In prehistoric times, about 1 A.D., the Woodland peoples began to move onto the Plains, living in semi-permanent villages. They sometimes interred their dead in small burial mounds. During the fifteenth century, a devastating and apparently long-continued drought forced the abandonment of the western-most villages. These people's migrations and turmoil during this period of drought have obscured their progression from prehistoric groups to historic tribes.[23]

Before they acquired horses and rifles, the Plains Indians did not grow crops, but instead depended on buffalo products for everything. Their shelters were large teepees covered with skillfully tanned hides. They sewed clothes and moccasins from the hides with buffalo sinew, made tools from the bones, and used buffalo chips as fuel for fires. The women cut the meat into strips and made jerky. When they moved to follow the hunt, they packed their belongings on the backs of large dogs or on travois, teepee poles pulled by a dog. The men were excellent with the bow and arrow. Using hand signs to communicate, the Plains Indians traded hides and meat for maize and cotton cloth with the Pueblo Indians. Some more easterly tribes established villages of earth lodges and supplemented their meat diet with maize, beans, and squash raised in river bottoms.[24]

The earliest Plains tribe to migrate to the Colorado area was the Apache, who inhabited the High Plains from the Black Hills to northern New Mexico and the Texas Panhandle from 1675-1725. During pre-horse times, they followed bison herds with the aid of dogs that transported their few belongings.[25] In the mid-seventeenth century, the Apache raided Spanish settlements in what is now New Mexico and carried off much livestock. The Pueblo Revolt of 1680 temporarily drove the Spaniards from New Mexico, placing even more livestock in Indian hands, accelerating the diffusion of horses. Plains Apache were so well supplied with horses by the end of the seventeenth century that they began trading them to other tribes. All of the Plains tribes south of the Platte probably had some knowledge of horses by 1700, and the use of horses had spread virtually to all of the Plains tribes by the mid-eighteenth century. The Apache, who historically were friendly with the Ute, today say they have oral traditions relating to the Garden of the Gods area. Most of the Apache were forced from this area by the arrival of other tribes.[26] One band of Apache joined the more powerful Kiowa tribe and managed to survive on the Plains as the Kiowa-Apaches.[27] The migration of other tribes that forced the Apache southward affected the relationships of the various peoples who passed along the Front Range of the Rockies and through the Garden of the Gods.

The Kiowa were related to the Tanoan-speaking pueblos of New Mexico who were primarily farmers, and perhaps were Puebloans who turned exclusively to hunting and went on extended bison hunting trips to the Southern Plains.[28] In the eighteenth century, the Kiowa were in south-central Montana at the headwaters of the Missouri, which was their aboriginal homeland. They were forced southward by the Cheyenne and Arapaho; however, at the Black Hills of South Dakota, the Kiowa were pushed back by the Sioux. The Kiowa eventually migrated to the southwestern Plains where they were nomadic bison hunters during the nineteenth century. In 1840, the Kiowa made peace with the Cheyenne, Arapaho, and Comanche, who along with the Kiowa-Apache, hoped to preserve the southern Plains for their use against settlement by Euro-Americans or the eastern Indian nations which were being moved to the West as a result of the Indian removal Act of 1830. However, in 1868 the Kiowas and Kiowa-Apache were sent to join the Comanche on Oklahoma reservations.[29]

Another tribe on the move was the Shoshone. They traveled from the Great Basin of Utah east through the Wyoming mountains to the northwestern Plains, then south into Colorado. The Shoshone did not go into the high country, instead, they hunted buffalo, antelope, and other game on the Plains. The Shoshone displaced the Apache, who moved on south.[30] Sometime in the seventeenth century, the Shoshone broke into two divisions, the Shoshone and the Comanche. A legend tells that the split occurred at the springs at Manitou Springs:

Shoshone 1884-1885
Denver Public Library

Many hundreds of winters ago, when the cottonwoods on the Big River were no higher than an arrow, and the red men, who hunted the buffalo on the plains—when, with hunting grounds and game of every kind in the greatest abundance, no nation dug up the hatchet with another because one of its hunters followed the game into their bounds, but, on the contrary, loaded for him his back with choice and fattest meat, and ever proffered the soothing pipe before the stranger, with well-filled belly, left the village,—it happened that two hunters of different nations met one day on a small rivulet, where both had repaired to quench their thirst. A little stream of water, rising from a spring on a rock within a few feet of the bank, trickled over it and fell splashing into the river. To this the hunters repaired; and while one sought the springs itself, where the water, cold and clear, reflected on its surface the image of the surrounding scenery, the other, tired by his exertions in the chase, threw himself at once to the ground and plunged his face into the running stream.

The latter had been unsuccessful in the chase, and perhaps his bad fortune and the sight of the fat deer, which the other hunter threw from his back before he drank at the crystal spring, caused a feeling of jealousy and ill-humor to take possession of his mind. The other, on the contrary, before he satisfied his thirst, raised in the hollow of his hand a portion of the water, and, lifting it towards the sun, reversed his hand and allowed it to fall upon the ground,—a libation to the Great Spirit who had vouchsafed him a successful hunt, and the blessing of the refreshing water with which he was about to quench his thirst.

Seeing this, and being reminded that he had neglected the usual offering only increased the feeling of envy and annoyance which the unsuccessful hunter permitted to get the mastery of his heart; and the Evil Spirit at that moment entered his body, his temper fairly flew away, and he sought some pretense by which to provoke a quarrel with the stranger at the spring.

"Why does a stranger," he asked, rising from the stream at the same time, "drink at the spring-head, when one to whom the fountain belongs contents himself with the water that runs from it?"

"The Great Spirit places the cool water at the spring," answered the other hunter, "that his children may drink it pure and undefiled. The running water is for the beasts which scour the plains. Au-sa-qua is a chief of the Shoshone; he drinks at the head water."

"The Shoshone is but a tribe of the Comanche," returned the other; "Waco-mish leads the grand nation. Why does a Shoshone dare to drink above him?"

"He has said it. The Shoshone drinks at the spring-head; Au-sa-qua is chief of his nation. The Comanche are brothers. Let them both drink of the same water."

"The Shoshone pays tribute to the Comanche. Waco-mish leads that nation to war. Waco-mish is chief of the Shoshone, as he is of his own people."

"Waco-mish lies; his tongue is forked like the rattlesnake's; his heart is black as the Misho-tunga [bad spirit]. When the Manitou made his children, whether Shoshone or Comanche, Arapahoe, Shi-an, or Pa-ne', he gave them buffalo to eat, and the pure water of the fountain to quench their thirst. He said not to one, Drink here, and to another, Drink there; but gave the crystal spring to all, that all might drink."

Waco-mish almost burst with rage. As the calm Shoshone stooped down to the spring to quench his thirst, the warrior of the Comanche suddenly threw himself *upon the kneeling hunter, and, forcing his head into the bubbling water, held him down with all his strength, until his victim no longer struggled, his stiffened limbs relaxed, and he fell forward over the spring, drowned and dead.*

Over the body stood the murderer, and a bitter remorse took possession of his mind. With hands clasped to his forehead, he stood transfixed with horror, intently gazing on his victim, whose head still remained immersed in the fountain. Mechanically he dragged the body a few paces from the water, which, as soon as the head of the dead Indian was withdrawn, the Comanche saw suddenly and strangely disturbed. Bubbles sprang up from the bottom, and rising to the surface, escaped in hissing gas. A thin vapory cloud arose, and gradually dissolving, displayed to the eyes of the trembling murderer the figure of an aged Indian, whose long, snowy hair and venerable beard, blown aside by a gentle air from his breast, discovered the well-known totem of the great Wan-kan-aga, the father of the Comanche and Shoshone nation.

Stretching out a war-club towards the frightened murderer, the figure thus addressed him:

"Accursed of my tribe! This day thou hast severed the link between the mightiest nations of the world, while the blood of the brave Shoshone cries to the Manitou for vengeance. May the water of thy tribe be rank and bitter in their throats." Thus saying, and swinging his ponderous war-club round his head, he dashed out the brains of the Comanche, who fell headlong into the spring, which, from that day to the present moment, remains rank and nauseous, so that not even when half dead with thirst, can one drink the foul water of that spring.

The good Wan-kan-aga, however, to perpetuate the memory of the Shoshone warrior, who was renowned in his tribe for valor and nobleness of heart, struck, with the same avenging club, a hard, flat rock which overhung the rivulet, just out of sight of this scene of blood; and forthwith the rock opened into a round, clear basin, which instantly filled with bubbling, sparkling water, than which no thirsty hunter ever drank a sweeter or a cooler draught.

Thus, the two springs remain, an everlasting memento of the foul murder of the brave Shoshone, and the stern justice of the good Wan-kan-aga; and from that day the two mighty tribes of the Shoshone and Comanche have remained severed and apart; although a long and bloody war followed the treacherous murder of the Shoshone chief, and many a scalp torn from the head of the Comanche paid the penalty of his death.[31]

After the split, the Shoshone were found primarily in Wyoming and Montana and the Comanche stayed in eastern Colorado where they competed with the Apache for hunting territory and in raids upon the Spanish and Pueblo settlements in New Mexico. They procured Spanish horses which they traded to their cousins the Shoshone, the Ute, and to the Pawnee and other tribes.[32] In 1760, the Ute and the Comanche had a disagreement and broke their alliance. For one hundred years they were at war, even though they were related people and spoke a similar language. In 1864, an attempt was made at peace, but somebody fired a shot so the Ute and the Comanche did not make peace until 1974 at the Ute - Comanche Peace Treaty Powwow.[33]

Another tribe claiming a connection to the Garden of the Gods is the Cheyenne. Before 1700, the Cheyenne moved to the Sheyenne River in North Dakota from the upper Mississippi Valley drainage where they had been displaced by the turmoil and upheavals of the seventeenth century. Well before 1800, the Cheyenne had become

Big Chiefs of the Cheyenne and Arapaho
Pikes Peak Library District, Langs

mounted bison hunters and they and the Arapaho and the Gros Ventres, all Algonquin-speaking people, came to dominate much of the western Plains between the Platte and the Arkansas, west of the Missouri.[34]

The Arapaho, who called themselves *Inunaina* or Our People, were true Plains Indians. They had originally lived in eastern North Dakota and Minnesota, north of the Cheyenne. They farmed and hunted bison and tended to ally with the Cheyenne. The Arapaho were among the first to move toward the west, entering the Great Plains in the mid-seventeenth century or earlier.[35] Moving southward with the Cheyenne in the late eighteenth and early nineteenth centuries, the Arapaho became primarily a hunting tribe, and lost the art of raising corn.[36] Between 1810 and 1820, the Arapaho moved between the North Platte and the Arkansas rivers, covering most of eastern Colorado and became traditional enemies of the Ute.[37] The Cheyenne and the Arapaho who frequently passed through the Garden of the Gods and up the Ute Trail to the summer hunting grounds in South Park, occasionally waged war on the Ute around the Garden of the Gods, and Ute Pass.

By 1835, the Arapaho had divided into northern and southern bands. The Northern Arapaho hunted buffalo along the North Platte River and the Southern, based near the Arkansas River, traded for horses with the Mexicans and Americans. The two sections of the tribe often visited each other or intermarried.[38] Many places in Colorado were considered sacred by the Arapaho. Waterfalls and other natural spectacles were signs of the presence of the Great Spirit. The Ute and Arapaho equally claimed the Boiling Springs. The site of present-day Manitou Springs was the scene of many skirmishes between the Ute and Arapaho and other Plains tribes that occasionally visited the Boiling Springs. According to Charles Marsh in *People of the Shining Mountains*, a small fortification was built by Indians for protection while using the waters. A circular stone barricade large enough to hold three or four men was used by many tribes. It is not clear who built these fortifications.[39]

The Pawnee, who had originally lived on the plains of east Texas, left their ancestral land sometime around the thirteenth century. Probably drought and increased population forced this move. They had been chiefly a farming people, and they sought out new farm land in the north. The Pawnee finally ended up on the rich banks of the Platte River and its tributaries more than three hundred miles from where they had started.[40] Although they lived in what is present day Nebraska and eastern Colorado, the Pawnee took hunting and trading trips south and west, coming in contact with the Ute along the Front Range. The Pawnee claim the Garden of the Gods as a religious site connected with their Morning Star tradition which required a young girl to be sacrificed to guarantee the success of their crops and the perpetuation of the universe. No evidence has been found to indicate any sacrifices took place in the Garden of the Gods area.

The Lakota Sioux are the last historical tribe to claim a connection to the Garden of the Gods. The Lakota sent scouts to look for a new territory for the tribe since eastern tribes were moving into their territory. Some of these scouts traveled as far south as the Garden of the Gods. Selo Black Crow, a Lakota medicine man, claims the Garden of the

Gods is one of the four power centers of the earth, and is the only place where all of the plants of their pharmacopoeia may be found. This indicates that even though the Lakota had little physical connection with the Garden of the Gods, it is a special or spiritual place for them. In 1994 the City of Colorado Springs Parks and Recreation Department commissioned a survey of the plant life in the Garden of the Gods. The result of the survey indicated more than 150 different plants could be found in the Park.

Plant Palate LIst for Garden of the Gods
Colorado Springs Park Department

PLANT PALATE LIST

Species name	Common name	Family	Bloom Date
Acer negunda	Box Elder	Acerac	
Yucca glauca	Yucca	Agavac	June
Allium cernuum	Nodding onion	Alliac	July
Allium textile	Wild onion		May
Amaranthus retroflexus	Rough pigweed	Amaran	
Rhus trilobata	Skunkbush	Anacar	May
Toxicodendron radicans	Poison ivy		
Apocynum androsaemifolium	Dogbane	Apocyn	July
Asclepias incarnata	Creeping milkweed	Asclep	
Asclepias incarnata	Creeping milkweed		June
Asparagus officinalis	Asparagus	Aspara	June
Cryptantha fendleri	Miner's candle	Boragi	June
Hackelia besseyi	False forget-me-not		June
Lappula redowski	Common stickseed		May
Lithospermum multiflorum	Many-flowered puccoon		June
Mertensia lanceolata	Narrow-leafed Mertensia		May
Onosmedium molle	False gromwell		
Coryphantha vivipara	Ball cactus	Cactac	June
Echinocereus viridiflorus	Hen and chickens		
Opuntia compressa	Prickly pear		July
Opuntia polycantha (?)	Starvation cactus		June
Cleome serrulata	Rocky Mountain bee plant	Cappar	Sept
Symphoricarpos occidentalis	Snowberry	Caprif	June
Saponaria officinalis	Soapwort	Caryop	July
Atriplex canescens	Four-winged saltbush	Chenop	June
Ceratoides lanata	Winterfat		July
Chenopodium album	Pigweed		Sept
Chenopodium botrys	Jerusalem oak		Sept
Chenopodium rubrum	Red goosefoot		Sept
Salsola kali	Russian thistle		
Tradescantia occidentalis	Spiderwort	Commel	June
Ambrosia tribida	Giant ragweed	Compos	Sept
Artemisia dracunculus	Wild tarragon		Augus
Artemisia filifolia	Wormwood		Septe
Artemisia frigida	Pasture sagebrush		Augus
Artemisia ludoviciana	Lance-leaf sage		Sept
Artium minus	Burdock		Septe
Aster porteri	Porter aster		Sept

Species name	Common name	Family	Bloom Date
Cirsium arvense	Canada thistle	Compos	July
Cirsium canescens	Thistle		June
Cirsium ochrocentrum	Thistle		June
Crysothamnus nauseosus	Rabbitbrush		
Dyssodia papposa	Fetid marigold		Septe
Erigeron divergens	Spreading fleabane		June,
Erigeron flagellaris	Whiplash erigeron		May,J
Grindelia squarrosa	Gumweed		July
Gutierrezia sarothrae	Snakeweed		July
Helianthus annuus	Sunflower		July
Helianthus pumilus	Perennial sunflower		
Heterotheca villosa	Goldeneye		July
Hymenoxys acaulis	Actinea		May
Hynmenopappus filifolia	Hymenopappus		June
Latuca tatarica pulchella	Lettuce		July
Liatris punctata	Blazing star		July
Lygodesmia juncea	Skeleton-weed		July
Machaeranthera bigelovii	Tansy aster		Augus
Machaeranthera pinnatifida	Machaeranthera		Augus
Ratibida columnifera	Prairie coneflower		July
Taraxacum officinale	Common dandelion		April
Townsendia grandiflora	Easter daisy		June
Townsendia hookeri	Easter daisy		April
Tragopogon dubius	Salsify		May,J
Verbesina enceliodes	Cowpen daisy		July
Convolvulus arvensis	Creeping Jenny	Convol	May,J
Sedum lanceolatum	Stonecrop	Crassu	
Camelina microcarpa	False flax	Crucif	May
Capsella bursa-pastoris	Sheperd's purse		May
Chorispora tenella	Blue mustard		May
Descurania sophia	Flixweed		May
Erysimum asperum	Western wallflower		June
Lepicium densiflorum	Peppergrass		June
Lesquerella ludoviciana	Bladder pod		May
Stanleya pinnata	Princes plume		June
Thlaspi arvense	Penny-cress		May
Euphorbia ecsula (?)	Spurge	Euphor	May/J
Quercus gambelii	Scrub oak	Fagace	June
Talinum parviflorum	Portulacaceae	Fame f	
Frasera speciosa	Monument plant	Gentia	June
Gentiana affinis (Nelson)	Prairie gentian		Sept
Erodium cicutarium	Storksbill	Gerani	March
Geranium bicknellii	Creeping Charlie		May,
Ribes aureum	Golden currant	Grossu	April

Species name	Common name	Family	Bloom Date
Juniperus monosperma	One-seed juniper	Pinace	March
Juniperus scopulorum	Rocky Mountain juniper		May
Pinus edulis	Pinyon pine		June
Pinus ponderosa	Ponderosa pine		June
Ipomopsis aggregata	Scarlet gilia	Polemo	
Erigonum umbellatum	Sulphur-flower	Polygo	
Eriogonum alatum	Winged eriogonum		July
Eriogonum jamesii var. flavescens	Bushy buckwheat		July
Eriogonum jamesii (flavescens)	False buckwheat		Septe
Rumex crispus	Curly dock		
Clematis hirsutissima	Sugarbowls	Ranunc	May
Clematis ligusticifolia	Virgin's bower		July
Delphinium ramosum	Larkspur		July
Pusatilla Patens	Pasque flower		April
Cercocarpus montanus	Mountain mahogany	Rosace	May
Crataegus crusgelli ?	Cockspur hawthorne		
Physocarpus monogynus	Ninebark		June
Potentilla pennsylvanica	Prairie potentilla		July
Prunum virginiana	Chokecherry		
Prunus americana	Wild plum		June
Prunus ?	Apple tree		April
Prunus ?	Apricot ?		
Rosa acicularis	Wood rose		June
Rosa woodsii	Wild rose		
Populus angustifolia	Narrow-leaved cottonwood	Salica	
Populus sargentii	Plains cottonwood		May
Salix amygdaloides	Peach-leaved willow		
Comandra umbellata	Bastard toadflax	Santal	June
Castilleja integra	Foothills paintbrush	Scroph	May
Linaria dalmatica macedonica	Butter-and-eggs		
Linaria vulgaris	Butter-and-eggs		
Penstamen angustifolius	Penstamen		May
Penstamen unilateralis (?)	Tall penstamen		July
Penstemon secundiflorus	Side-bells penstemon		May
Verbascum thapsus	Mullein		July
Physalis lobata	Purple-flowered ground cherry	Solana	June
Ulmus pumila	Siberian elm	Ulmace	April
Conium maculatum	Poison hemlock	Umbell	July
Cymopterus acaulis	Cymopterus		April
Cymopterus montanus	Cymopterus		April
Verbena bracteata	Prostrate vervain	Verben	

Species name	Common name	Family	Bloom Date
Ribes cereum	Wax current	Grossu	May
Ribes inerme	Common gooseberry		May
Iris missouriensis	Rocky mountain iris	Iridac	May/J
Hedeoma drummondii	Pennyroyal	Labiat	
Nepeta cataria	Catnip		July
Scutellaria brittonii	Skullcap		
Astragalus bisulcatus	Two-grooved milkvetch	Legumi	June
Astragalus drummondii	Drummond milkvetch		May
Astragalus shortianum	Early purple milkvetch		May
Dalea pupurea	Purple prairie clover		July
Medicago sativa	Alfalfa		July
Melilotus officinalis	Sweetclover		May,J
Mililotus alba	White sweetclover		July
Oxytropis lambertii	Lambert loco		June
Psoralea tenuiflora	Scurf pea		July
Psoralea tenuiflora	Pea		July
Robinia neomexicana	New Mexican locust		June
Thermopsis rhombifolia	Golden banner		April
Trifolium fragiferum	Strawberry clover		July
Vicia americana	American vetch		May
Calochortus gunnisonii	Mariposa lily	Liliac	June
Leucocrinum montanum	Sand lily		May
Smilacina stellata	False Solomon's seal		May
Zygadenus veneosus	Death camas		
Linum lewisii	Wild flax	Linace	
Mentzelia speciosa	Many-flowered evening star	Loasac	June
Sphaeralcea coccinea	Copper mallow	Malvac	June
Oxybaphus linearis	Narrow-leafed umbrellawort	Nyctag	July
Fraxinus pennsylvanica	Ash	Oleace	
Calylophus serrulata	Evening primrose	Onagra	
Gaura coccinea	Scarlet gaura		June
Gaura parviflora	Gaura		July
Oenothera coronopifolia	Cutleaf evening primrose		May,J
Oenothera caespitosa	White stemless evening primrose	Onogra	
Orobanche fasciculata	Clustered cancer-root	Oroban	
Argemone hispida	Prickly Poppy	Papave	June
Abies concolor	White fir	Pinace	April
Juniperus communis alpina	Common juniper		

Indians traveling to the Garden of the Gods, the Boiling Springs, or heading west over Ute Pass took two main routes. Those coming from the north or east from the Plains made a trail through Garden Ranch and Templeton Gap, then crossed Monument Creek about a mile above Colorado Springs, possibly where Garden of the Gods Road is today. The trail followed a ridge to the mesa, then southwest over the mesa and across Camp Creek, passing through the Garden of the Gods. The trail then went down to Fountain Creek about one mile west of Colorado City and joined another trail from the southeast, up the east side of Fountain Creek. Coming from the southeast, the Indians made a trail that followed the east side of Fountain Creek from the Arkansas River, crossing Monument Creek just below where an Artificial Ice Plant was located in Colorado Springs. From this point, the trail ran along the north side of Fountain Creek to a point just west of Colorado City where it crossed to the south side of the creek and led up to the Boiling Springs. From here the trail went up Ruxton Creek for a few hundred yards, crossed over to the west side, then up the creek to a point just below the Colorado Midland Railway Bridge. The trail continued westward up a long ravine to its head, then in the same direction near the heads of the ravines running into Fountain Creek. A quarter to a half mile south of Fountain Creek it continued west for two miles or more. The trail finally came back down to Fountain Creek below Cascade Canyon and from there led up along the creek to its head, where it branched off in various directions. This is the famous old Ute Pass Trail used by

Ute Trail, 1810-1820
Pikes Peak Library District, Poley

various tribes of Indians for hundreds of years before the European discovery of America. This trail is one of the oldest documented routes of any of the Native Americans. It was used for many generations by explorers, hunters, trappers, and Indians until the white settlers came, and even after that by occasional war-parties until the Indians were driven from their traditional homes.[41]

Navajo Spring, Manitou Springs Heritage Center, Jackson, Hayden Survey 1870

The use of the Indian trail by early explorers led them to the Boiling Springs at Manitou. Dr. Edwin James, the botanist and historian of Colonel Stephen Long's expedition, who visited the Pike's Peak region in 1820, stated in his journal: "A large and much frequented road passes the springs [Manitou] and enters the mountains running to the north of the high peak."[42] In Rufus B. Sage's book *Rocky Mountain Life,* published in 1846, he states that, "the Fontaine qui Bouille Creek is derived from two singular springs situated within a few yards of each other at the creek's head...The Arapaho regard this phenomenon with awe and venerate it as a manifestation of the immediate presence of the Great Spirit. They call it Medicine Fountain and seldom neglect to bestow their gifts upon it whenever an opportunity is presented. These offerings usually consist of robes, blankets, arrows, bows, knives, beads, moccasins, etc. which they either throw into the water, or hang upon the surrounding trees."[43]

Dr. James noticed in the bottom of the spring a great number of "beads and other small articles of Indian adornment."[44] According to George Frederick Ruxton in *Ruxton of the Rockies*, to the Indians, especially the Arapaho, "...the 'medicine' waters of these fountains are the abode of a spirit who breathes through the transparent water, and thus, by his exhalations, causes the perturbation of its surface. The Arapaho, especially, attribute to this water god the power of ordaining the success or failure of their war expeditions; and as their braves passed often by the springs when in search of their hereditary enemies the Ute, they never failed to bestow their offerings upon the water sprite, in order to propitiate the Manitou of the fountain, and ensure a fortunate issue to their 'path of war.' " At the time of his visit to the springs, Ruxton noted the basin was filled with beads, wampum, pieces of red cloth, and knives, while the surrounding trees were hung with strips of deerskin, cloth, and moccasins. The sign around the spring plainly showed that a war dance had been executed by Arapaho braves on their way to war with the Ute.[45]

Battle Between Ute and Comanche
Denver Public Library, Drannan

Warfare was frequent between the Ute occupying the Garden of the Gods and the other tribes who wanted to go up the pass. Irving Howbert, one of Colorado Springs' first and most famous pioneers maintains in his book *The Indians of the Pike's Peak Region* that the Ute had forts, "...anyone climbing to the top of a high sandstone ridge back of the United States Reduction Works at Colorado City might have seen numerous circular places of defense built of loose stone, to a height of four or five feet, and large enough to hold three or four men comfortably."[46]

Rock Fort,
Author's Collection, Evans

Several locations have been found in the Garden of the Gods which seem to resemble such advantageous lookout points; because of their height, a considerable area could have been kept under surveillance. Numerous stone chips have been found in such spots which lead to the theory that the sentinels kept busy with the making of arrowheads while they were on guard duty.[47] In addition to the forts, the Indians hid from their enemies in a cave located below North Gateway Rock.

Paul Cimino, caretaker of the Garden of the Gods for over thirty years, has a collection of more than 300 perfect arrowheads, all found on the Garden of the Gods grounds. He discovered some campsites and that is where he found many of the arrow heads. He believes there were three main campsites at one time, and that the Ute used the Garden of the Gods for a battleground with Plains Indians.[48] The Ute usually spent their summers in the South Park area, so the Plains tribes would travel through the Garden of the Gods on their way up Ute Pass to South Park in order to raid the Ute and try to steal their horses.

At one time, it was believed evidence of a battle in the Garden of the Gods had been discovered.

Building a Teepee,
1890-1900
Pikes Peak Library District, Poley

On November 1, 1925, a tourist taking a photograph of an odd boulder noticed a skull in a fissure after he had walked about 200 yards west of the Ridge Road, near the Gateway rocks. Three-fourths of the surface of the skull was exposed and the joints were visible. According to newspaper reports, upon excavation an entire skeleton was found in the flexed (fetal) position with an Indian arrow between the third and fourth left ribs. Near the left hand of the Indian was a bone bow, and in the grave were many worn and decomposed bone arrows. Coroner Swann declared the remains to be those of an Arapaho Indian who had been killed at least 100 years prior to this discovery. He was thought to have been killed in a war between the Ute and the Arapaho tribes. Authorities of Indian lore felt he was probably high in the ranks of the Arapaho tribe, as he was an exceptionally large Indian and had been buried, apparently with honors. His rank was also evident from the arrows found with him.[49]

John Deal, Buckskin Charlie, and Ocapoor
Pikes Peak Library District, Poley

Controversy over the origins of this skeleton swirled for sixteen years. Shortly after the discovery, Dr. Edgar L. Hewett, director of the American School of Research at Santa Fe, New Mexico, was in town to address the Winter Night Club on prehistoric civilization of the southwest. After a brief survey of the skeleton and its burial place, Dr. Hewett stated he believed the find to be one of great scientific and historical interest, particularly as it was possible that other bodies could be buried there as well.[50]

A report in the *Colorado Springs Gazette* on Friday, November 27, 1925 maintains that Dr. J. A. Jeancon, curator of the State Historical and Natural History Society of Denver, prepared a report for the Colorado Springs City Council in which he stated that "the laws of natural science show beyond contradiction that the skeleton could not have been buried there prior to a century ago at the most... He believes the formation contiguous to the arroyo is of soft sandstone that is readily eroded. Since the skeleton was buried less than eight feet underground, the natural processes of erosion easily could have crumbled that depth of the sepulcher to eight feet within the past century." The reporter goes on to state that "Throughout his report he [Jeancon] takes sharp issue with the opinions advanced by Dr. Edgar Hewitt, director the American School for Research and Museum at Santa Fe, N.M., who stated the bones were those of a Pawnee Indian woman and were of great antiquity."[51]

The Pawnee do not have a tradition of flexed burial. The Comanche, on the other hand, would take the body before the warmth had gone and bend the knees upon the chest and the head forward to the knees. The body was then bound in this position with a rope. The corpse was bathed, the face overlaid with vermilion, and the eyes sealed with red clay.... The preferred burial place was a natural cave, crevice, or deep wash among the rocks of the highest possible peak, or in the head of a canyon, preferably to the west of the lodge of the departed.[52]

E. B. Renaud, Professor of Anthropology at the University of Denver wrote a paper in June of 1941 entitled "Western and Southwestern Indian Skulls." In this paper he reports on the ethnic origin of the same skull found in the Garden of the Gods:

> *Early in November, 1927 [sic], an Indian skeleton was discovered buried in the Garden of the Gods, between Colorado Springs and Manitou, Colorado. The late Mr. J. A. Jeancon, then Curator of Archaeology at the State Museum at Denver, and Dr. Edgar L. Hewett, Director of the School of American Research and Museum at Santa Fe, New Mexico, dug it out and examined it first. Later I was asked to see the skeleton and help identify it. The following is a report based on the measurements I took of the skull and the long bones, the study of the indices obtained and the comparison with the ethnic elements of the region.*
>
> *It is reported that the grave contained a skeleton in the flexed position but no archaeological remains of any kind were found, either costume, implements, pottery or the like, to help identify period, culture, or tribe....*[The implements reportedly found with the skeleton were apparently lost or stolen before Renaud did his study.]
> *From all appearances the skull is that of a female....*[53]

Renaud goes on to say that the measurements of the long bones (leg and arm) indicate the woman was about 62.03 inches tall, far shorter than the typical Arapaho and Cheyenne. As a result of his measurements and close inspection of the skeleton, Renaud concludes that this woman was of an earlier time than the Plains Indians.[54] The relatively short stature of the skeleton could place it as either Comanche or Ute. According to Renaud's report, the skull showed "a mixture [of genetic influences] with more or less localized characteristics showing the respective influence of the various elements on different parts of the skull."[55]

Finally, according to an article in the *Sunday Gazette and Telegraph* dated February 15, 1931, the skeletal remains were returned to Colorado Springs and interred in the Garden of the Gods next to an old Indian trail which was used by the Plains Indians to traverse the park on their way to the springs at Manitou. Both the grave and the trail were marked by large boulders.[56]

Endnotes:

1. Dr. James Goss, ed., "Proceedings Garden of the Gods American Indian Workshop" (Colorado Springs, Colorado: October, 1994),76.
2. *Ibid.*, 66.
3. Tomas E. Ross and Tyrel G. Moore, ed., *A Cultural Geography of North American Indians* (Boulder: Westview Press, 1987), 50.
4. *Ibid.*, 49.
5. *Ibid.*, 54.
6. J. Donald Hughs, *American Indians in Colorado* (Boulder, Colorado: Pruett Publishing Co., 1977), 27.
7. Pettit, 4.
8. Woodard, 12.
9. Charles S. Marsh, *People of the Shining Mountains* (Boulder: Pruett Publishing Co. 1926), 56.
10. Pettit, 5.
11. Hughs, 26.
12. "Uto-Aztecan Languages." (http://en.wikipedia.org/wiki/Uto-Azetecan_Languages. 2008), 4-5.
13. Alden Naranjo, Interview by author, E-mail, Tuesday, April 8, 1997, Monument, Colorado.
14. "Introduction to Ute Tribal History," 2.
15. Pettit., 15-17.
16. *Ibid.*, 28-30.
17. "Introduction to Ute Tribal History," 2.
18. Pettit, 21.
19. *Ibid.*, 21-22.
20. *Ibid.*, 101.
21. *Ibid.*, 91-92.
22. Goss, 77-78.
23. William W. Newcomb Jr., *North American Indians: An Anthropological Perspective* (Pacific Palisades: Goodyear Publishing Co. Inc., 1974), 81-83.
24. Hughes, 23-24.
25. *Ibid.*, 83.
26. *Ibid.*, 85-86.

27. *Ibid.*, 87.
28. Alice Beck Kehoe, *North American Indians: A Comprehensive Account* (Englewood Cliffs: Prentice Hall, Inc., 1981), 278.
29. *Ibid.*, 288-289.
30. *Ibid.*, 277.
31. Irving Howbert, *Indians of the Pikes Peak Region* (Glorietta: The Rio Grande Press, Inc., 1970), 38-42.
32. *Ibid.*, 287-288.
33. Goss, 102.
34. *Ibid.*, 86.
35. Hughes, 35.
36. Kehoe, 280-281.
37. Hughes, 35-36.
38. *Ibid.*, 36.
39. Marsh, 168.
40. Maxwell, 158.
41. Howbert, *Indians*, 27-28.
42. *Ibid.*, 28-29.
43. *Ibid.*, 30-31.
44. LeRoy R. Hafen, ed., "Rufus B. Sage: Letters and Scenes in the Rocky Mountains," *Far West and Rockies Series*, 1820-1875, Vol. V (Glendale, California: Arthur H. Clark Co., 1956), 75.
45. LeRoy R. Hafen, ed., *Ruxton of the Rockies* (Norman: University of Oklahoma Press, 1950), 237.
46. Woodard, 14.
47. *Ibid.*, 14.
48. Debra Skodack, "Guardian of the Garden" (*Gazette Telegraph*, Sunday, July 13, 1980), 5 BB.
49. "Skeleton Found In Garden of the Gods," *The Colorado Prospector* (1927), 6-7.
50. "Skeleton Was Indian Woman, Experts Aver," *Colorado Springs Gazette* (Colorado Springs: Friday, 6 November 1925), 9.
51. "Experts Differ Over Antiquity of Skeleton," *Colorado Springs Gazette* (Colorado Springs: Friday, 27 November 1925), 12.
52. Ernest Wallace, and E. Adamson Hoebel, T*he Comanches, Lords of the South Plains* (Norman: University of Oklahoma Press, 1952), 150.
53. E. B. Renaud, "Western and Southwestern Indian Skulls" *Anthropological Series*, First Paper (Denver: University of Denver Department of Anthropology, 1941), 12-13.
54. *Ibid.*, 17-18.
55. *Ibid.*, 18.
56. "Indian Trail and Grave In Garden of Gods Now Marked By Huge Boulders," *Sunday Gazette and Telegraph* (Colorado Springs: 15 February 1931), 6.

Explorers, Trappers, and Traders

During the sixteenth century, the pinnacles and spires of the Garden of the Gods witnessed the gradual development of another type of human interaction. The sole-custodianship of the Indians was on the wane and the ownership of the White man was about to begin.

Beginning in 1535 with Cabeza de Vaca's exploration of the North American southwest, Spain claimed a stretch of country of indefinite width east of the Rocky Mountains extending from Mexico north to the Yellowstone River.[1] In 1763 Spain gained official ownership of the land from the Mississippi River to the Pacific Ocean through the Treaty of Paris, then in 1800, through the secret treaty of San Ildefonso, Spain ceded the region to France. In 1803, Napoleon sold it to the United States as part of the Louisiana Purchase.

During the period of Spanish ownership, many people explored the region from Santa Fe to the Platte River. The earliest recorded Spanish expedition to the Pikes Peak area was in 1598. Juan de Onate, a prominent Spaniard of the time and a relative of Cortez and Montezuma, attempted a large expedition into the northern country for exploration and possible colonization. This expedition moved up the Rio Grande Valley into the San Luis Park region of Colorado. About thirty miles north of Santa Fe, Onate founded the town of San Gabriel, the second in the territory. A short time later he sent his nephew Juan de Zaldivar, with a company of cavaliers, farther into the interior. He progressed along the foothills nearly to the site of Denver. In 1601, Onate started another expedition northeastward, both to continue Zaldivar's search and to learn more of the ill-fated 1593 expedition of Francisco Leyva de Bonilla and Antonio Gutierrez de Humana.[2] They journeyed for over three months and came as far north as Denver, then turned eastward into eastern Kansas and reportedly went as far as the Missouri River, either in Kansas or Nebraska.[3]

Records of New Mexico show a military expedition traveled through the country north of the Arkansas River and east of the Mountains in 1719 and another led by Lieutenant-

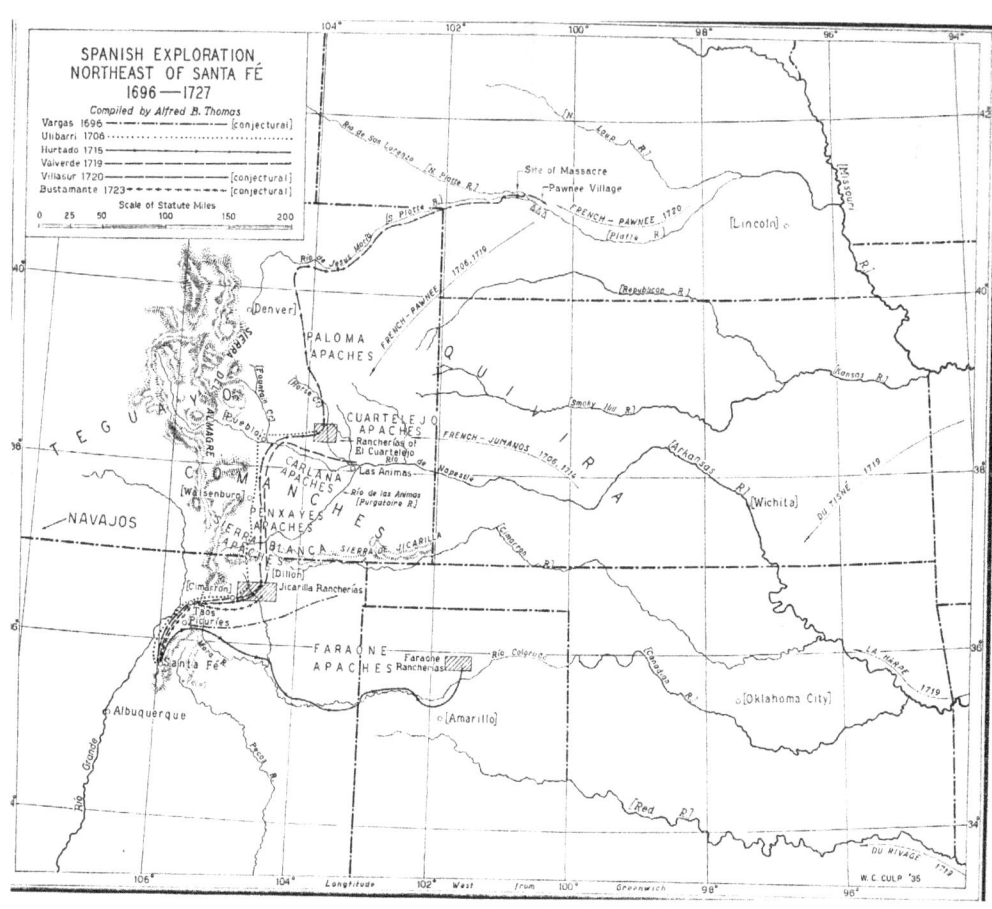

Spanish Exploration North of Santa Fe, 1696-1732
Map, Colorado College

General Pedro de Villasur ventured northward the following year. It is very possible that one or both of these expeditions passed over the future town site of Colorado Springs. While digging a cellar at 529 East Pike's Peak Avenue in June 1894, workmen uncovered, about six feet below the surface, an iron box four by six inches in size, which was rusted to the point of disintegration. This iron box held a crucifix, attached to which was a Maltese cross having a small erect cross as a pendant. The two crosses were made of dark colored stone resembling chalcedony, while the crucifix was of solid hammered brass. These articles were hand made, showing the hammer marks and filed edges. None of the three bore any inscription or date, but evidently had been lost by a priest accompanying one of the Spanish expeditions.[4] According to Irving Howbert, these articles disappeared before they could be put some place secure. Many other Spanish parties visited this region between 1720 and 1819, when Spain finally waived its right to this territory.[5] The line between New Spain and the United States had been in dispute until the Adams-Onis Treaty established the official border.[6]

Any explorers who came through the area of the present Colorado Springs would have seen the Garden of the Gods and very likely went through it on their way up Ute Pass or to visit the boiling springs. Jonathan Carver, an Englishman who traveled through the Far West from 1766 to 1768, reported a legend that the early Spaniards climbed Pikes Peak and that early explorers found an altar of rough granite at the summit. This altar has, of course, long since disappeared.[7]

The French names in the Pikes Peak region originated with French trappers from Canada who drifted through the area during the eighteenth and early nineteenth centuries, penetrating Ute Pass to South Park's lush grasslands. They left their traces in place names around Pikes Peak. The Fountain Creek was first called the Fountain qui Bouille, or the fountain that boils, because of its turbulence as well as because of the bubbling springs in the vicinity.[8]

In 1739, two Frenchmen, the Mallet brothers explored a route from a point on the Platte southwestward to the Arkansas River, then went on down to New Mexico. They were the first to document reaching New Mexico from the Northeast.[9] In 1751, the French traders Jean Chapuis and Louis Feuilli secured permission to engage in trade with New Mexico from the commandant

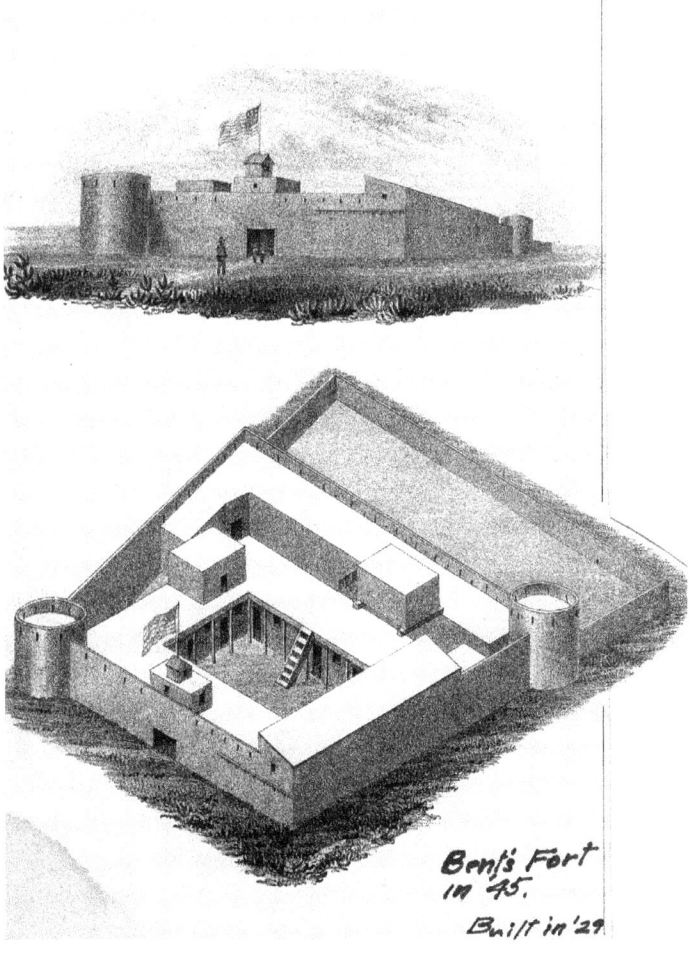

Bent's Fort
Colorado College

Bent's Fort in '45.
Built in '29.

at Michilimakinac in Canada and from Captain Benoit de Sainte Claire at the fort at la Chartre. They left in early October, taking a route which first took them to Fort Cavagnille, then along the Platte River, then southwest. Upon their arrival in New Mexico, a Mexican official named Luis Febre interrogated them, then sent them to Mexico where they were further questioned before being sent on to Spain and prison.[10] This was typical of the way the French and later American trappers and traders were received upon arrival in New Mexico with their goods. Consequently, many trappers chose not to trap in the mountains of Colorado because it was too difficult to get their furs to market. In 1833, two Saint Louis brothers, William and Charles Bent, and a Missourian named Ceran St. Vrain opened a trading post fort on the Arkansas River. Bent's fort gave the mountain men a place to swap their beaver pelts for supplies[11] and had the effect of opening up the fur trade in the Colorado Rockies, which brought more trappers and traders through Ute Pass and into contact with the Garden of the Gods area. Joe Doyle, a trapper who worked for Bent and St. Vrain Co. in 1842, along with Dick Wooton and William Kroenig, owned the Soda Springs on the Fountaine qui Boille in 1859. They planned to develop the springs into a spa they called the Saratoga of the West.[12]

Jesus Silva and "Uncle Dick" Wooton
Denver Public Library, Sargent

One of the most famous explorers to reach the Pikes Peak region was Zebulon Pike, who set out in the summer of 1805 for the West. He traveled up the Fountain from the Arkansas with the purpose of climbing the Grand Peak. On November 21, 1805, he succeeded in climbing a nearby mountain, thinking it was the main peak until he saw it was at least another day's march to the base of the Grand Peak. Pike stated he believed no human could ascend to its summit. On his descent, he observed several Tetau (Ute) camps. On December 3, Pike and others measured the altitude of the peak and then proceeded south towards Canyon City.[13] Zebulon Pike never climbed the peak, nor made it to the north side of it, thus one may presume he never saw the boiling springs nor the Garden of the Gods.

The next expedition of significance was that of Major S. H. Long. In 1820, Major Long headed west, following the North Platte River. According to Captain John R. Bell, the official journalist of the expedition, they secured the services of two Frenchmen to act as guides, "It was with great difficulty, the Commanding officer could engage two Frenchmen, traders residing with the Indians [Pawnee Republic Village on the Loupe Fork of the Platte River], to accompany our party as far as the Arkansas River—liberal offers of reward had no effect and it was not until he threatened to report them to the government and have them removed from the Villages, that they agreed to go. Mr. Bijeau [Bijou or Bissonet] was hired as Indian interpreter and Guide and Mr. Ladeau [Abraham Ledoux] as Pawnee interpreter. Each was paid one dollar per day."[14] Both men were residing permanently among the Pawnee and had been repeatedly on the headwaters of the Platte and Arkansas for the purposes of hunting and trapping beaver. Bijeau was partially acquainted with several Indian languages. He had been with trappers Choteau and DeMun in 1815 on a trapping and trading venture in the southwest when they ranged widely on the upper waters of the Arkansas and Platte. Bijeau had trapped beaver in the mountains for about six years.[15]

ZEBULON MONTGOMERY PIKE
1779 — 1813
Discoverer of Pikes Peak — 1806
Sesquicentennial 1806 - 1956

Zebulon Pike
Pikes Peak
Library District

Major Stephen H. Long
Colorado College

The Long expedition headed south from Platte Canyon on July 9 and ascended Willow Creek to its source. They crossed a ridge to Plum Creek and followed it some distance before making camp. Pikes Peak first came into view on the ninth of July from the top of a mesa. On the tenth they discovered a rock formation and named it Castle Rock. Some confusion exists as to whether this is Castle Rock in the town of that name or Elephant Rock in Palmer Lake. According to Dr. James, "One of these singular hills, of which Mr. Seymour has preserved a sketch, was called the Castle Rock, on account of its striking resemblance to a work of art. It has columns, and porticos, and arches, and, when seen from a distance, has an astonishingly regular and artificial appearance."[16]

James also goes on to describe a lake sitting upon the divide between the waters that flow into the Platte and the waters that flow into the Arkansas. This description and the sketch of the rock closely resemble Elephant Rock and Palmer Lake. The party continued south, fording Monument Creek, and toward evening of the eleventh they discovered they had passed the base of Pikes Peak. They stopped at this point so they could climb Pikes Peak and determine its altitude.[17]

Bijeau told the party about the boiling springs, Fountaine qui Bouille, and guided some of the men to them. Captain Bell wrote of the excursion: "It is a custom with the wandering bands of Indians that travel this region of country, when passing these springs, to make offerings to the Great Spirit, by casting ornaments of beads, shells, etc. into them, attended with a sort of ceremony or religious form. Some of these articles of sacrifice was obtained by Lieut. Swift from the springs & brought in. Bijeau informs us that French traders when in this country, examine these springs & take from them articles thus religiously deposited by the nations and trade them off to the same people again for skins (Bell 136-137)."[18]

The members of this expedition also commented on a "large and frequented road" which passed the springs into the mountains. This road was the old trail through the Ute Pass.[19] Previous to 1840, Pikes Peak was known as James Peak. Major S. H. Long gave it this name in honor of Dr. James who is supposed to have been the first white man to ascend it. After about 1840, this name was gradually dropped and Pike's Peak was substituted in honor of Zebulon Pike.[20]

The next United States expedition, and the first of strictly military character to march to the Rockies, was that of Colonel Henry Dodge. The purpose of this visit was peace among the whites and with the Indians. In 1835, the First Regiment of United States Dragoons set out for the West. On the afternoon of July 9th, the expedition entered Colorado at the northeastern corner of the state, following the right bank of the South Platte. They went up the east side of the river, crossing Denver's site and preceded almost to the mouth of Platte Canyon. They then went up Plum Creek, over the ridge to Monument Creek, down the Monument to the Fontaine qui Bouille, and traveled down the stream to a place about fifteen miles from the mountains. The expedition continued southwest to the Arkansas, and after a stop at Bent's Fort, they continued down this stream into Kansas.[21]

Rufus B. Sage visited the Pikes Peak area in 1842 and described the scenes surrounding Manitou Springs:

> ...And there are other scenes adjoining this that demand a passing notice. A few miles above Fountaine qui Bouit, and running parallel with the eastern base of the mountain range, several hundred yards removed from it, a wall of coarse, red granite towers to a varied height of from fifty to three hundred feet.
>
> This wall is formed of an immense strata planted vertically and not exceeding eight feet in thickness, with frequent openings - so arranged as to describe a complete line.
>
> The soil in which they appear is of a reddish loam, almost entirely destitute of other rock, even to their very base.
>
> This mural tier is isolated and occupies its prairie site in silent majesty, as if to guard the approach to the stupendous monuments of nature's handiwork that form the back-ground, disclosing itself to the beholder for a distance of over thirty miles.[22]

Sage states in his book *Rocky Mountain Life* that he found the climate mild, the wildflowers enchanting, the game abundant, the scenery heavenly. He thought it little wonder the Indians revered the place as home to the Good Spirit.[23]

During the summer of 1843, John C. Fremont, with Kit Carson as guide, went from the mouth of the Fountain northward along the stream to the springs of Manitou.[24] His description of the trip is as follows:

John Fremont
Colorado Springs
Pioneer Museum

Went up the "Fountaine-qui-bouit" river to sample the springs—camped there, then descended the river in order to reach the eastern fork, which I proposed to ascend. The left bank of the river here is very much broken. There is a handsome little bottom on the right and both banks are exceedingly picturesque—strata of red rock, in nearly perpendicular walls, crossing the valley from north to south. About three miles below the springs, on the right bank of the river, is a nearly perpendicular limestone rock, presenting a uniformly unbroken surface, twenty to forty feet high, containing very great numbers of large univalve shells, which appears to belong to the genus inoceramus... To the westward, was another stratum of limestone, containing fossil shells of a different character; and still higher up on the stream were parallel strata, consisting of a compact somewhat crystalline limestone, and argillaceous bituminous limestone in thin layers.

During the morning, we travelled up the eastern fork of the Fountaine-qui-bouit river, our road being roughened by frequent deep gullies timbered with pine and halted to noon on a small branch of this stream, timbered principally with the narrow-leaved cottonwood (populus augustifolia) called by the Canadians laird amere. On a hill near by, were two remarkable columns of a grayish-white conglomerate rock, one of which was about twenty-feet high, and two feet in diameter. They are surmounted by slabs of a dark ferruginous conglomerate, forming black caps, and adding very much to their columnar effect at a distance. This rock is very destructible by the action of the weather, and the hill, of which they formerly constituted a part, is entirely abraded.[25]

James Hall, a paleontologist traveling with Fremont gives the following description of the area of the Garden of the Gods:

Longitude 105, latitude 39—The specimens from this locality are a somewhat porous, light-colored limestone, tough and fine grained. One or two fragments of fossils from this locality still indicate the Cretaceous period; but the absence of any perfect specimens must deter a positive opinion upon the precise age of the formation. One specimen, however, from its form, markings, and fibrous structure, I have referred to the genus noceramus.

It is evident, from facts presented, that little of important geological change is observed in traveling over this distance of 7 degrees of longitude. But at what depths beneath the surface the country is underlain by this formation, I have no data for deciding. Its importance, however, must not be overlooked. A calcareous formation of this extent is of the greatest advantage to a country; and the economical facilities hence afforded in agriculture, and the uses of civilized life, cannot be overstated.

The whole formation of this region is probably, with some variations, an extension of that which prevails through Louisiana, Arkansas, and Missouri.

The strata at the locality last mentioned are represented as being vertical, standing against the eastern slope of the Rocky Mountains, immediately below Pikes Peak.[26]

Kit Carson
Denver Public Library,
Barry

Another early visitor to the Garden of the Gods and the Fountaine Qui Bouille was a British explorer, Lt. George Frederick Ruxton in 1847. In the early spring, Ruxton went up the Fountaine qui Bouille to hunt and to pasture his animals in the mountains. In his journal he describes the Garden of the Gods, "On the western side the rugged mountains frown overhead, and rugged canyons filled with pine and cedar grape into the plain. At the head of the valley, the ground is much broken up into gullies and ravines where it enters the mountain spurs, with tops of pine and cedar scattered here and there, and masses of rock tossed about in wild confusion. On entering the broken ground, the creek turns more to the westward, and passes by two remarkable buttes of red conglomerate which appear at a distance like tablets cut in the mountainside."[27]

Ruxton goes on to describe his journey up the Ute Pass to the springs:

I followed a very good lodge-pole trail, which struck the creek before entering the broken ground. Here the valley narrowed considerably and turning an angle with the creek. I was at once shut in by mountains and elevated ridges, which rose on each side of the stream. This was now a rapid torrent, tumbling over rocks and stones, and fringed with oak and a shrubbery of brush. A few miles on, the canyon opened out into a little shelving grade; and on the right bank of the stream, and raised several feet above it, was a flat white rock in which was a round hole, where one of the celebrated springs hissed and bubbled with its escaping gas. I had been cautioned against drinking this, being directed to follow the stream a few yards to another, which is the true soda spring. The escape of gas in this was much stronger than in the other and was similar to water boiling smartly.

[The taste] was equal to the very best soda water, but possesses that fresh, natural flavour, which manufactured water cannot import.[28]

Trapper Jacob Spaulding and two companions camped in the Garden of the Gods during the winter of 1848-1849 after failing to find a pass through the mountains on their trek to California. Four feet of snow fell the first week of November, two mules had to be destroyed to keep the animals from starving; the other pair were fed on the bark and tree branches. Spaulding discovered a cavern hidden inside North Gateway Rock, 200 feet long, five to fifteen feet wide, and nearly 100 feet high. A stream of cold water trickled down from the ceiling. He found tracks of bobcat and mountain lion on the floor. He discovered the cave was also a good echo chamber. Spaulding moved on in the spring, but he returned in 1862.[29]

Another white man to frequent the area of the Garden of the Gods was Thomas Fitzpatrick, Indian Agent and Mountain Man, who traveled up the Arkansas to Pueblo, then up Fountain Creek and crossed the Palmer Divide to the South Platte drainage in 1853 to give the Arapaho and Cheyenne their allowance.[30] The era of explorers, trappers and traders gradually gave way to the gold seekers, settlers, and entrepreneurs after the discovery of gold in 1859. The cry Pikes Peak or Bust drew the hopeful to the Pikes Peak region and the Garden of the Gods.

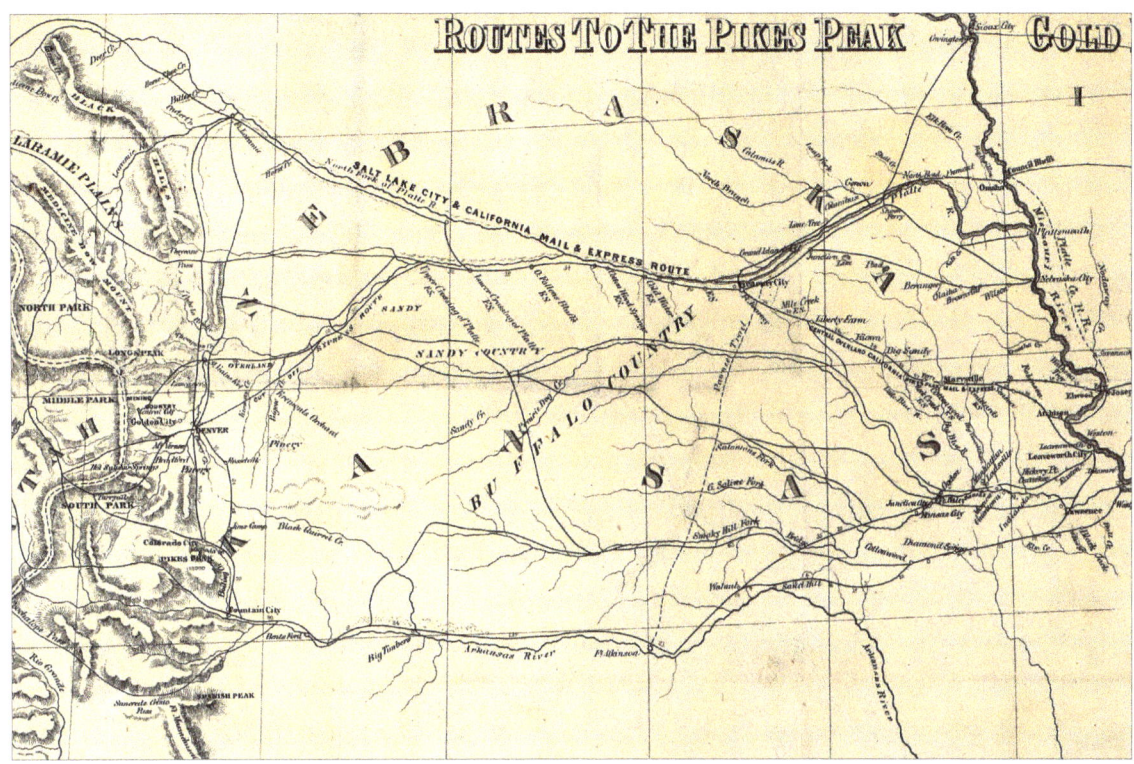

Routes to the Pikes Peak Gold Region
Map, Colorado College

Endnotes:

1. Irving Howbert, *Memories of a Lifetime in the Pikes Peak Region* (Glorietta: The Rio Grande Press, Inc., 1925), 85.
2. "New Mexico: The Onate Expeditions," www.wisconsinhistory.org, 10/2/2008.
3. Wilbur Fisk Stone, ed., *History of Colorado: Volume I* (Chicago: The S. J. Clarke Publishing Co., 1918), 25-26.
4. Howbert, *Memories*, 85-86.
5. *Ibid.*, 85.
6. Sons of Dewitt Colony Texas, *New Spain-Index* (McKeehan, Wallace L., 1997-2007), www.tamu.edu/ccbn/dewitt/Spain3.htm.
7. John Fetler, *The Pikes Peak People: The Story of America's Most Popular Mountain* (Caldwell, Idaho: The Caxton Printers, Ltd., 1966), 26.
8. *Ibid.*, 49.
9. LeRoy R. Hafen, ed., *The Mountain Men and the Fur Trade of the Far West, Vol. III* (Glendale, California: Arthur H. Clark Co., 1966), 15-16.
10. *Ibid.*, 21-26.
11. Roger G. Kennedy, ed., *The Smithsonian Guide to Historic America: The Rocky Mountain States* (New York: Stewart, Tabori, and Chang, 1989), 79.

12. *Missouri Democrat*, November 20, 1859 (Colorado Historical Society, MSS XXV), 56.
13. Zebulon M. Pike, *The Southwestern Expedition of Zebulon M. Pike* (Chicago: The Lakeside Press, 1925), 78-81.
14. Hafen, *The Mountain Men*. 29.
15. *Ibid.*, 29.
16. Stone, 54.
17. *Ibid.,* 54.
18. Hafen, *The Mountain Men*. 30.
19. *Ibid.*, 55.
20. Irving Howbert, *Indians of the Pikes Peak Region*. (Glorietta: The Rio Grande Press, Inc., 1970), 49.
21. Stone, 80.
22. Howbert, *Indians*. 32.
23. Richard Gehling, and Mary Ann, *Man in the Garden of the Gods* (Woodland Park: Mountain Automation Corp., 1996), 4-5.
24. Stone, 59.
25. John Charles Fremont, "Report of the Exploring Expedition to the Rocky Mountains," *March of America Facsimile Series Number 79* (Ann Arbor: University Microfilms, Inc., A Subsidiary of Xerox Corporation, 1966), 118 [174].
26. *Ibid.*, 296 [174].
27. Leroy Hafen, *Ruxton of the Rockies*. (Norman: University of Oklahoma Press, 1950), 235.
28. *Ibid.*, 235-236.
29. Gehling, 5.
30. Carl Ubbelohde, ed., *A Colorado Reader* (Boulder: Pruett Press, Inc., 1967), 82-83.

Miners, Settlers, and Entrepreneurs

As evidenced by the names carved in the rocks in the Garden of the Gods, a large number of goldseekers passed through the area in their search. Some 118 names were carved into the rocks of the Garden of the Gods between 1731 and 1890. Calvin Clark, a Pikes Peak goldseeker, noted in August 1859 that numerous names had already been cut into White Rock. More than two dozen names still remain from the gold rush days, but Clark's name has vanished.

Autograph Rock
Pikes Peak
Library District,
Zelley

Names found on the major rocks in the Garden: North and South Gateway, White Rock, Gray Rock, Sleeping Indian, and Sentinel Rock. Underlined names have been identified.

H.O. Acom
F.L. Austin
<u>I.L. Avery 1858</u>
H.C. Baldman
D.A. Ban
___. H. Bates
A.W. Baxter
C. B. Bert
E. Blackstone
F.E. Blurk
<u>O. Brooks 1859</u>
O.C. Cabington
Calfas
A.B. Cooley
<u>J. Coplen 1860</u>
<u>W.H. Conner</u>
S.E. Conner
G.A. Copley
C.A. Crandall
J.E. Crantham
D.A. Deeme
J. Demps
C.L. Dewitt
F.B. Dexter
<u>D.P. Drake</u>
Cy M. Duncan 1872 MO
H.F. Dunkel
A. Dunlop
<u>J.B. Ely</u>
E. Elliott
O.B. Emerson
A.L. Fawld
J.H. Ferguson
Reif Flarentin
<u>Mrs. Lou Frost 1870</u>
W.H. Gabrief
<u>John M. Good</u>
H.C. Graf
Gubbison
L.D. Halden
<u>Wm. Hartley 1858</u>
Dorthy Hanse<u>n</u>
<u>T. Hartman</u>
Haswell 1871
Heinzma

Chas. He<u>d</u>ges
<u>J.A. Hope</u>
I. Hottet
E. Hullard 1873
<u>E.S. Imes</u>
F.H. Jacob 1877
<u>M.M. Jewett</u>
<u>A.H. Jones</u>
<u>A. Kendall</u>
J.L. Kermer
W. L.L. Ketner 1731*
U.E. Kinnel
C.H. Kirkeride
F. Kocherhons 1785*
M. Kohn
H. Lape 1807 MO*
R.C. Lashley
<u>W. Lierd '70</u>
Lisele
A. Lonert 1878
A. Lord
A. L. Low
Ch. McCoke
<u>H.W. McKnight</u>
W.B. McLewis Augus 1, 185_
McNally
S.T. M Man<u>n</u>
P. Menole
<u>D.W. Mills 1866</u>
W. Moise
C.L. Moses
J.H. Nance
<u>W. Nelson 1866</u>
G.H. Nolan
John Nolan
<u>C.E. Palmer</u>
H. Pelvik
L.V. Pinke 1879
Pinif
Jhn Potts
J. Ramer
R.C. Ransom
Fron Rinigton June 12, 1874
A.P. Robinson
<u>J. Robins</u>

A.B. Sane<u>s</u> (g) MO
<u>A.B. Sanford 1880</u>
G. P. Schield
J.R. Seato 18_8
Schadow
H.O. Shulte
D. Smith
G.F. Smoes
<u>W.A. Stephens</u>
A.A. Tillarson
Timi
A. Ulm
<u>A. S. Voorhees</u>
B.F. Wadsworth
D. Wanore
F.J. Wendlin
Chas Weng
<u>J.C. Williamson</u>
Zan Wilson
W.H. Winter
<u>W.B., F, W.H. Wiswall 1870</u>
J.R. Witt
<u>A.C. Wright 1858</u>
M.D. Wright
M. Woods
<u>Geo Young</u>
H. Yount
<u>T.C. Yutsey 1860</u>

Names Carved on Rocks, Pioneer Museum, Garden of the Gods Folder

In 1935, the cave in the North Gateway Rock was rediscovered at the request of an old man who brought his grandchildren to the Garden of the Gods to see the cave he had played in as a youth. At that time park officials knew nothing of the cave, but their interest was aroused. John A. MacDougal, superintendent of the Palmer Park Civilian Conservation Corps (CCC) camp, and enrollees at work in the Garden of the Gods discovered and opened the cavern. Inside they found where many of the earliest of the pioneers had carved their names with dates upon the walls. According to an article in the *Sunday Gazette and Telegraph*, October 13, 1935, it was impossible to identify the names with the dates in many instances because the carvings were so numerous and so irregularly carved upon the walls. In one place, there was a date of 1858, which was before Colorado City was founded. Unfortunately, many of the names were illegible.[1] Carving names in the rock was a tradition which continued until the Garden became a city park. Now carving anything into the rocks is forbidden in order to protect the historic names and dates and to protect the rocks themselves.[2]

More than thirty of the carvers have been identified. Above a small cave on the west side of White Rock are two names within a few inches of one another, WLL Ketner 1731 and F Kocherhons 1785. These dates may or may not be authentic. Parties of French trappers and

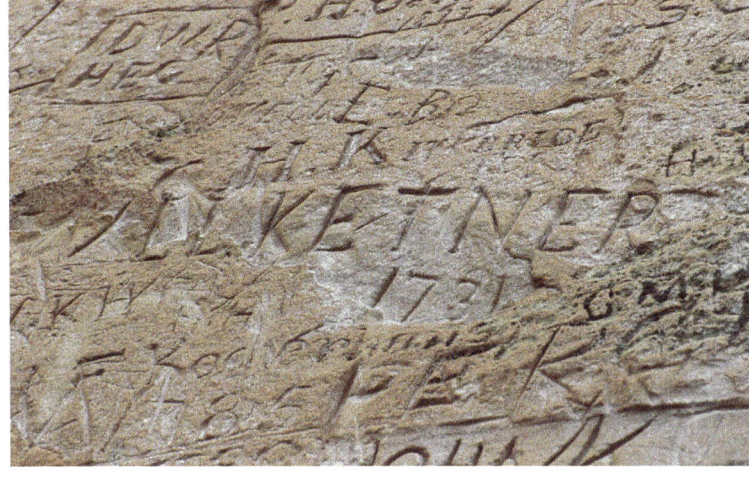

Ketner and Kocherhons
Author's Collection,
Evans

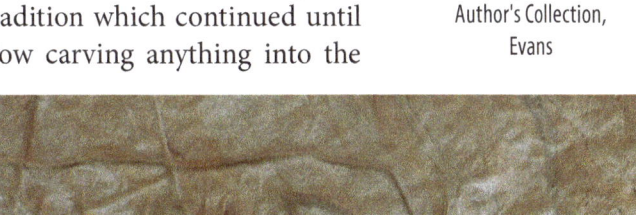

H. Lapey, 1807 MO
Author's Collection,
Evans

traders were in the area in the mid to late 1700s, and Ketner and Kocherhons were possibly fur trappers. Another early name and date is found above the same cave on the west side of White Rock. The carving H LAPe 1807 MO is difficult to read because of the presence of later carvings. If this name is authentic, it probably belonged to a French trader or trapper from St. Louis.

A Dutch goldseeker from Iowa named Huiscamp had to spend the winter of 1850-1851 camped in the Garden of the Gods with the Ute, probably because of a broken wagon axle. In the spring he followed his Ute friends up Ute Pass into South Park. His descendants returned to Colorado Springs to open Alexander Aircraft and Film Company in 1919.[3]

Alexander Airfield 1920s
Pikes Peak Library District

On the south side of a ten-foot by fifteen-foot rock just to the southeast of North Gateway Rock is the name A.C. Wright 1858. The date is in a star composed of two equilateral triangles. Andrew C. Wright, better known by his nickname Jack, was born in New York City but grew up in Natrick, Massachusetts. He moved to Lawrence, Kansas in 1855 at the age of twenty. He was a member of the Lawrence Party which camped on Camp Creek from July 8, to August 10, 1858. Wright remained with the Lawrence Party during their move north to the Russell Diggings at the mouth of Cherry Creek in September of 1858. By October 1, he had moved up the South Platte to Henderson Island. There he went into camp with squawmen (men who had married into an Indian tribe, in this case, probably Pawnee) William McGaa, Bill Roland, and Jim Saunder.

When word came of the organization of St. Charles (later Denver City) by the Lawrence Party, Wright and his new companions decided to return to the mouth of Cherry Creek. On the west bank of the creek, they built a two-story log cabin. The town of Auraria was soon laid out around it. By the spring of 1860, Andrew Wright was back in the Pikes Peak region, probably living in Colorado City. In May or June of that year, he and Jersey Hinman decided to lay claim to Jimmy Camp, the famous stopover on the Old Trappers' Trail some fifteen miles east of the Garden of the Gods. The two men laid out a foundation of logs just down the hill from the famous Jimmy Camp springs. They stayed only a day and a half before deciding to abandon their claim.[4] Wright returned to Lawrence and married Miss Cordelia E. Ricker. He and his new bride soon returned to Colorado City. After a couple of other moves, Wright and his wife and two children settled in Denver where he opened Denver's first livery stable with George Estabrook.

Wright 1858
Author's Collection, Evans

Another member of the Lawrence Party to carve his name on the rocks was William Hartley (Wm Hartley 1858 south side of the same rock as Wright). Hartley was with the Lawrence Party during the move north to the Russell Diggings at the mouth of Cherry Creek in September of 1858. He was a civil engineer and surveyor. T.C. Dickson described what followed:

> *Shortly after this a party of us, including William Hartley, who was a surveyor and had his instruments with him, decided to lay out a town on the east side of the Platte River, about five miles from the mouth of Cherry Creek, which we did, and called it 'Montana.' About a week after this Charles Nichols, Adna French, John A. Churchill, Frank M. Cobb, William Hartley, W.M. Smith, William McGaa, John S. Smith and myself, nine persons in all, who were interested in the Montana townsite, came to the conclusion that we had made a mistake in laying out a town five miles away from the regular trail... So we went down to the mouth of Cherry Creek, surveyed and laid out the town of St. Charles, Consisting of 640 acres lying east of the Platte River and Cherry Creek. This was on the 24th of September, 1858.*[5]

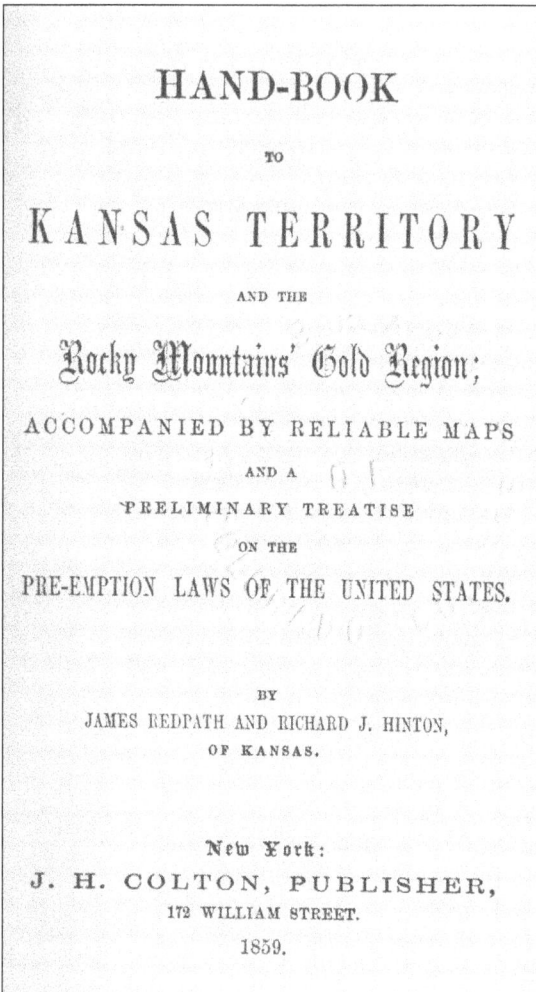

Hinton Handbook
Colorado College

After laying out the town of St. Charles, the town company, including William Hartley, decided to spend the winter in Lawrence and return the following spring. While they were gone, the townsite was jumped by the Larimer Party from Leavenworth, Kansas. The area was re-surveyed and renamed Denver City. Sometime after their return to Lawrence, William Hartley and T.C. Dickson collaborated on a guidebook to the new goldfields. It was advertised as "A descriptive guide book with tables showing camping places; distance between the same, on the various routes, with a comprehensive and reliable map of the newly discovered gold regions…." This was to be among the first of many guidebooks written for sale to the Fifty-Niners. A copy of it could be obtained by mailing one dollar to Wm Hartley and Co., St. Louis, Mo.

Augustus Voorhees, also a member of the Lawrence Party, carved his name "A S Voorhees" on the walls of the great cavern inside North Gateway Rock. Voorhees, of German-Dutch ancestry, was born in New York State on December 14, 1828. He moved to Wisconsin when he was eighteen and to Kansas when he was twenty-eight, where he farmed and did some coal mining.

Voorhees was working at the Coal Bank, nine miles northeast of Burlingame, Kansas when he heard that the Lawrence Party had started to Pikes Peak in late May of 1858. He raced thirty-five miles on foot to overtake the train, finally catching the ten wagons at Bluff Creek. Voorhees was the only member of the Lawrence Party to keep a daily diary. His entries began on May 31 and continued until July 12, the date when the goldseekers moved to Jimmy Camp from the Garden of the Gods. Here his diary comes to an abrupt end, and it is not known if he kept any further records of the trip. The original diary is in possession of the State Historical Society of Colorado.

While encamped on Camp Creek, Voorhees and two others climbed to the top of Pikes Peak. Of this climb Voorhees wrote in his diary:

July 8 - Miller cooked four days provisions to go to Pikes Peak.

July 9 - We started in the morning, followed up the creek valley three miles, and found three soda fountains or boilling fountains. They are quite sour and resemble Congress water. They boill up very strong. We also found a log shanty there. We climbed up the mountain all day, found some hard climbing and large rocks. I killed one sage hen which we roasted for supper. We made our camp between two rocks and built a large fire. We had a light shower in the night. They had a heavy hail storm at the camp. We crept under the rocks.

July 10 - Started early and went to the foot of the main peak and left our blankets, and went on up. We had a heavy hail storm, which kept us three hours under the rock. The mountain was covered with hail. We got to the top at three o'clock, but it was so cloudy we could not see the country beyond. We cut our names on a stick and p (ut) it in a pyramid of stones that we piled up. The top is level twenty-five or thirty acres, nothing but small rock tumbled together. We started down at four o'clock, it was so cloudy that we could see nothing below us. We could not find the way to our packs, and we stoped at the first timber which was pick pine, and found a rock to shelter us from the wind and rain. We build up a great fire and stoped for the night. It was so cold we did not sleep much without our blankets. We were one mile from the top. It blew very hard all night.

July 11 - The fog blew off this morning so that we could find our packs, and we started for home. The mountains are covered with spruce, the fire has burnt most of them dead. The red raspberey and strawbereys are quite thick. Some of them are ripe and some are in blossom. The fog was so thick that we could not see much as we came down. We found some quarts rock and some crystal quarts. We came into camp at four o'clock (Voorhees Diary).[6]

Voorhees stayed with the Lawrence Party during the trip south to the Old Spanish diggings in Grayback Gulch, then north to the new diggings on Cherry Creek. In the fall of 1858, he returned to Kansas with other members of the Lawrence Party. When the Civil War broke out, he enlisted in the Fifth Kansas Cavalry, serving for three years. Later he returned to Wisconsin, then in 1890 returned to Kansas. Despite all this movement, he found time to get married and raise a family of seven. He died in Kansas in 1905.[7]

Trail up Pikes Peak
Denver Public
Library District,
Duham Brothers

Summit of Pikes Peak
Denver Public
Library DIstrict,
Gurnsey

Returning from the Summit
Pikes Peak Library District,
Thurlow

The carving M.M. Jewett Aug. 7O is found on the south side of North Gateway Rock. Marshall M. Jewett was a member of the Larimer Party of goldseekers which was organized at Leavenworth in September of 1858. They left for the goldfields on October 1 with eight wagons, six months supply of provisions, and thirty-two men. The Larimer Party arrived in Auraria on November 16, 1858, where General Larimer and his men formed the Denver City Company and adopted a constitution. Their town was named in honor of James W. Denver, the governor of Kansas.

Marshall M. Jewett seems also to have been an original member of the Colorado City Town Company organized the following year. He was likely one of the dozen men present when the town company was first organized. The site of the meeting was the office of Richard Whitsitt on Larimer Street in Denver City on August 11, 1859. According to M. S. Beach, the company had been organized "for the purpose of laying out a town at the entrance of the Ute Pass at the base of Pikes Peak... So after the company was organized, M. S. Beach and R. E. Cable were delegated to proceed on horseback to the new townsite and locate it as Colorado City." Two days later, on August 13, 1859, a sign was erected southeast of the Garden of the Gods, "claiming 1280 acres, extending a mile wide and two miles long from Camp Creek to Monument Creek." [8]

M.M. Jewett
Author's Collection,
Evans

Driving the First Stake
Pikes Peak
Library DIstrict

In late February 1860, Marshall Jewett sold one half of his interest to Joseph M. McCubbin for $150 cash or merchandise. On April 26, 1860, Jewett entered into contract with John Gerrish, E. Cobb, and J. Lauber to purchase their original interest in the townsite for $425. On July 13, 1860, Jewett sold the other half of his original interest to the same McCubbin for an extra $400, thus realizing a profit of $125.[9]

The name J Coplen 1860 is carved just above Marshall Jewett's on the south side of North Gateway Rock. Eighteen-year-old John Coplen came to Colorado City from Ohio in 1860. He made the trip with his parents, William and Ruth Coplen (Ruth Coplen was said to be a cousin to President James A. Garfield). John was the eldest son in a family of fourteen, several of whom did not survive childhood. He fought with Company G, Third Regiment, Colorado Volunteer Cavalry in the Indian wars of 1864. This Regiment fought under the command of Colonel Chivington at Sand Creek. After mustering out on December 19, 1864, Coplen filed a land claim in Black Forest. Reportedly in later years, John Coplen invented the Coplen Concentrator for concentrating ores.[10]

On the southeast side of North Gateway Rock is the name E.S. Imes. Elsworth Imes, along with two brothers and six sisters, came to the Pikes Peak region in 1865 with his parents, Moses J. and Mary (Davis) Imes. They apparently came for the mother's health, who, in fact, made the trip sitting on a rocking chair inside the covered wagon. In the company were also two uncles, Lew and Blackhawk Davis. Elsworth's father and his older brother William both filed for El Paso County land at the Denver Land Office on December 14, 1869. According to the El Paso County 1867 Tax Schedule, Elsworth and perhaps his brother William owned E. W. & Co., a sawmill in 1867.

On North Gateway Rock is carved John M Good. John Good was born in Alsace-Lorraine, France in 1835. He came to the United States when he was nineteen and settled in Akron, Ohio. At the age of twenty-four, Good joined the great Pikes Peak Gold Rush to Cherry Creek. At Denver City in 1859, he opened one of the first general merchandising stores in the new settlement. He freighted his goods overland from the Missouri Valley, making as many as sixteen trips over the next several years. By April of 1860, John Good had entered in partnerships with Fred Solomon and Charles Endlich for the purpose of manufacturing lager beer under the name of Solomon & Co.

In 1862, John Good returned to Indiana and married Rosalie M. Wagner. They had four children during the next eight years. By 1870, John Good had established the Rocky Mountain Brewing Company. Living in his household was one Philip Zang from Hesse, Germany, a worker in the brewery. Some years later, John Good sold the brewing company to Philip Zang. After leaving the brewery business, John Good invested heavily in mining and banking, in railroad and real estate. He accumulated a large fortune and was twice appointed treasurer of the city of Denver. He died in Denver on November 24, 1928, at the age of eighty-four.[11]

Beer Depot
Denver Public Library

General Albert H. Jones carved his name high above the little cave on the west side of White Rock in 1869. (A H Jones 1869 St. Louis, Mo.) Jones was born in Newark, New Jersey in 1839. During the Civil War, he fought with the Hawkins' Souaves, Company B, Ninth New York Infantry. Shortly before the end of the war, he was transferred to a gunboat and was so badly crushed by a falling mast that he was honorably discharged with the rank of corporal.

A. H. Jones 1888
St. Louis, MO
Author's Collection, Evans

After Jones had recovered from his wounds, he came west. He arrived in Denver in 1866 and opened a wholesale liquor store on Sixteenth and Market Streets. In 1876, he married and had a son, also named Albert. In 1878, Jones organized the Chaffee Light Artillery Company. For a time he served as captain of the governor's guard, and for nearly fifteen years he acted as brigadier general and inspector general of the Colorado National Guard. In 1890, General Jones was appointed United States Marshall for Colorado. After retiring from that office, he pursued his mining interest until his death on November 16, 1910 in Denver.[12]

One of the few early women to carve their names on the rocks was Mrs. Louisa B. Frost (Mrs. Lou Frost 1870 on the walls of the large cave inside of North Gateway Rock). Louisa was born in 1846 in Missouri. When she was nineteen, her parents decided to move to the Colorado Territory. On April 7, 1863, they crossed the Missouri River at St. Joseph and organized with one hundred other wagons into a train under a Captain Cross. The members of this train were setting out for Oregon, California, Idaho, and Colorado. Several incidents of that journey across the plains were to remain with Louisa the rest of her life:

While we were encamped on the Little Blue, my chum and I went to the river to get a pail of water. We two used to do so many risky things! We saw a log across the river and from it a path leading up the hill on the opposite side. We set the pail down and went exploring. It was a winding trail up thru the new green of early spring. Soon we saw a house. The door was open and we went in. To our great surprise, an Indian squaw was lying ill. She kept motioning to her mouth and pointing to a vessel nearby. We guessed she wanted a drink, so we hunted for a cup and gave her one. She seemed grateful as much as she could make us understand. While we were there the door was darkened suddenly and we looked around to see a large Indian man. He turned out to be her husband. In her tongue she evidently told him what we had done and he seemed pleased. We took them to be civilized Indians of the Pawnee tribes, but we got out quickly and started back. My chum's father was hunting for us by this time and met us on the trail. What a scolding! We were always doing things we shouldn't.

After strenuous traveling for several days the captain told us we could stop and rest for a day or so. We went down to the river to do some washing. Just as we had the clothes well soaked with water, we were told to pack everything at once and come back to camp as some Indians were there and acting very strange... Some wore buckskin but most of them were almost nude. They watched us silently for quite a while.

Our men were all on guard and well armed so they didn't try anything... The captain figured they might be intending to bring more men later on for an attack, consequently when the Indians were well out of sight, he ordered us to pack up and cross the river for a better position....

Guards listened to every movement on the prairie that night. If the Indians were coming back it was expected at dawn as that was the favorite time for them to attack. So long before daylight, our camp was up. But we were not molested that time or any other time, and made it to Colorado without getting scalped.[13]

By 1870, Louisa was married to Edward W. Frost, former government scout, Indian fighter, and veteran of the Civil War. They lived for a while at St. Vrain's fort, now Longmont. The Census of 1870 shows them living at Virginia in El Paso County, probably located in Black Forest. Edward was twenty-four and a farmer. Louisa was twenty-three and had a son Edward Jr., then one year old. The Frost family later moved to Colorado Springs, where they raised three sons. Edward became an alderman, water commissioner, and later health commissioner. He died in 1914 at the age of seventy-nine. Louisa survived him for twenty-two years. She died in 1936 at the age of ninety.[14]

Historian Albert B. Sanford, who had been born in Denver City in 1862, carved his name A B Sanford 1880 high above and to the right of the little cave on the west side of White Rock. Albert was eighteen when he engraved his name in the Garden of the Gods.

Half Century Club 1922, Louisa Frost fifth woman from right.
Pioneer Museum

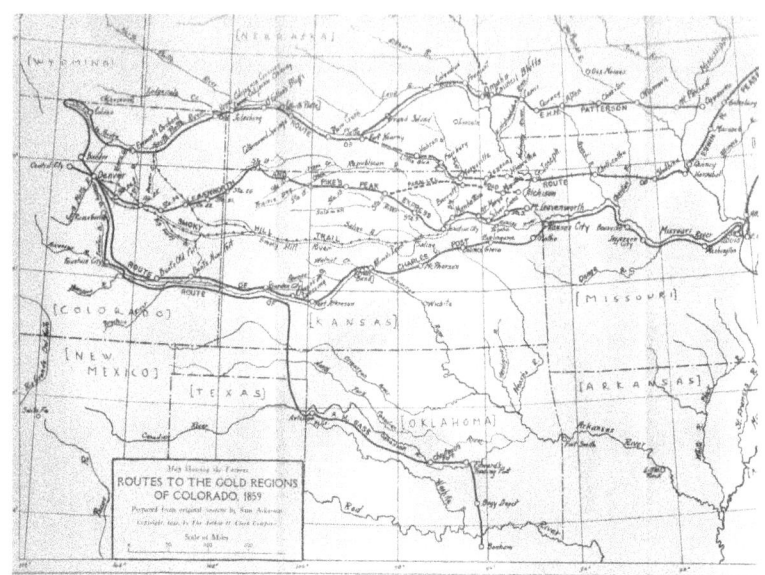

Overland Routes to the Gold Regions of Colorado.
Colorado College

In later years Albert Sanford wrote a series of articles for The Trail magazine. In 1925, he was appointed Curator of History for the State of Colorado, largely through the efforts of the Sons of the Pioneers.[15]

The name W. Lierd '70 is carved in the large cave inside of North Gateway Rock. This name was clearly visible in 1935 when the cave was rediscovered. William A Lierd was born October 30, 1846 in Burlington, Iowa. He first came to El Paso County from Chicago in August of 1870. He lived for many years in Monument where he kept a grocery store. In later years, William and his wife Emma moved to Colorado Springs and settled there. Both were members of the El Paso County Pioneer Society.[16]

Many of these hardy pioneers who left their names carved in the rocks of the Garden of the Gods came here seeking gold in the Pikes Peak region. Instead, some of them found their fortunes by trading with others who came out to dig gold. The first attempt at settlement near the Garden of the Gods was made by a party of goldseekers from Lawrence, Kansas. Times were hard and a get-rich-quick attitude prevailed. Word of gold in the Pike's Peak region spread throughout the East. A group of forty goldseekers organized for safe conduct to the mountains. James H. Holmes and his wife and Robert Middleton and his wife and child joined the group. They left

Lawrence, Kansas on May 19, 1858 with eleven wagons drawn by oxen. Their route took them over the Santa Fe Trail to the Arkansas River, then north to the mouth of the Fontaine qui Bouille, then up to the Pike's Peak area. On July 8, 1858, the wagon train approached the Garden of the Gods. They camped just east of the red rocks alongside a little stream which came to be known as Camp Creek. It was a stream flowing out of Glen Eyrie and down the valley to Fountain Creek at Colorado City.

For more than a month they lived out of their wagons, panning for gold, eating, smoking, and snoozing. They carved finger rings and pipes out of soft gypsum that covered the backside of White Rock. Others spent their time playing euchre and cribbage. They climbed the rock formations and explored many caves, carving their names into the soft sandstone.[17] Several members of the party climbed Pikes Peak, Mrs. Holmes included. Julia Archebald Holmes, later known as "The Bloomer Girl," was the first white woman to climb Pike's Peak. They stayed on the summit for two days and Mrs. Holmes wrote letters to the eastern press dated July 1858.

In Camp
Colorado College

Soon after the party established camp along Camp Creek, a season of heavy and continuous rain set in. They were flooded out and moved into the cave in North Gateway Rock. Visitors to this shelter many years later said that the smoke stains of their fires still blackened the walls and that the names of two or three of them cut into the stone were still visible. Some of these people were instrumental in the starting of Colorado City.[18]

The first settlement in the Pikes Peak Region was on the Fontaine qui Bouille, a short distance above the mouth of Monument

Getting Supper
in the Rockies
Photo, Colorado College

Creek. During the winter of 1858-1859, a company named the "El Paso Town Company" chose a site for a town which was laid out as El Paso City. This was located at an Indian Trail in the gateway to the Ute pass through to the South Park, thus the name El Paso. Little is known of the character of this town organization or the names of those who were active in its formation. The town of El Paso lay within the present boundaries of Colorado City. El Paso experienced a short life and was succeded during the following summer and fall by Colorado City.[19]

In 1859, after an announcement had been made of the discovery of gold in the Terryall District in South Park, Melancthon S. Beach decided to establish a trading post at the entrance to Ute Pass through which there was a ready route to the new gold region. Beach organized the Colorado City Town Company in Denver. With Rufus E. Cable as his guide, they made the trip down from Denver on horseback and proceeded immediately to stake out Colorado City (West Colorado Springs) on August 14, 1859, two hours before another group came down from Denver with the same idea in mind. The two-hour leeway gave them precedence.[20]

The town site lay on the Fountaine qui Bouille from a point near the gypsum bluffs above the mouth of Camp Creek toward the mouth of Monument Creek, encompassing 1280 acres of land two miles long and one mile wide. Beach and Cable christened their town Colorado City after the nearby red rocks - *colorado* is Spanish for red.[21] They then staked adjoining claims along Camp Creek in the shadow of the red rocks.

The second party moved over to stake out what was once called Roswell in the north part of Colorado Springs, (Main Street west of Tremont, between Fillmore and Van Buren Streets today). They named their site "Red Rocks," from the red colored rocks at the Gateway entrance to the Garden of the Gods which they could see from their location.[22]

It was at this point that the Garden of the Gods received its name. In 1858, the Garden was called the Red Rock Corral. However, "Beach and Cable rode over to inspect the Garden of the Gods, and as they gazed in wonder and amazement at the grand sight which greeted them, Beach said, 'What a place this would be for a beer garden!' But Cable apparently had been filled with reverence and gratitude for the handiwork of the Supreme Architect, for he replied with feeling: 'It is a fit place for a 'Garden of the Gods,' let that be its name!'"[23]

In a letter to the Boston *Watchman and Reflector*, written in 1870, Mr. Lewis N. Tappan relates virtually the same story except that he states, "The writer, in company with

Colorado City, 1860
Photo, Colorado College

three other gentlemen..." He credits Cable with the same reply, so there were really four in the party that participated in the naming.[24] Irving Howbert in *Memories of a Lifetime in the Pikes Peak Region*, relates a similar version of the story:

> *According to M. S. Beach, principal founder of Colorado City, the Garden of the Gods was so named by R. E. Cable, a lawyer from Kansas City, here on a visit in August 1859. Mr. Beach states: 'Cable was then a young and poetic man, and when we visited the site together I suggested that it would be a capital place for a beer garden when the country grew up. He exclaimed, 'Beer garden! Why, it's a fit place for the gods to assemble. We will call it the Garden of the Gods' and by that name it was known from that date. No one should endeavor to take that honor from Rufus Cable.' Although this name was universally recognized from that time, the locality was more often spoken of by the early settlers as 'The Red Rocks.'* [25]

In early days references were made to a couple of Gardens of the Gods. The First Garden of the Gods, or the Big Garden being Garden of the Gods, and the Second Garden of the Gods, or the Little Garden being Glen Eyrie. Governor William Gilpin, the first Governor of the Territory of Colorado, is generally credited with

Half Century Club,
Melancthon S. Beach
2nd row left
Photo, Pioneer Museum

Early Tourists at the Gateway
Photo, Old Colorado City Historical Society

trying to give the Garden the name of The Garden of the Immortal God, but this name was never accepted.[26]

The unusual formations of the Garden of the Gods inspired people to write glorious descriptions to their friends and relatives back east. The morning edition of the Missouri Democrat dated Tuesday, March 20, 1860 ran the following letter by Professor Goldrick (the editor of the *Rocky Mountain Herald* in 1924). He was writing from "Colorado City, Foot of Pikes Peak, El Paso County, Jefferson Territory, March 1, 1860," and speaking of the situation of Colorado City and Pikes Peak, "…its huge and lofty castled domes and towers of variegated rocks and sandstone pyramids, which stand around the outskirts, and especially adjoining the 'Garden of the Gods', like towering sentinels who have held their position there since Noah's flood, these ponderous pedestals and eruptive entablatures, terrace above terrace, and these hillocks of conglomerations and petrifactions, lying, as it were, in a regular confusion, are a curiosity and a study for the gravest and greatest geologists…"[27]

Opie Read, a famous author of the time, wrote an article for the New Year Number of the *Gazette Telegraph*, January 1, 1905:

> *...in this mysterious place, what jolts there are to the imagination! In exaggeration of fancy, it is the dream of a giant. In its myriad of animal shapes it is the menagerie of eternity, petrified; and but for its grandeur it would be a verbal storm. It is earth's most weird emotion caught and stilled on the verge of terrific outbreak. Milton's Satan would have claimed it for his theater. After beholding the rest of the world, Adam would have stood in surprise. It is sublimity's cloister. From it Carlyle might have caught the ruggedness of his style. Profanity would term it the brick kiln of hell. Humor would call it the delirium tremens of nature. But truth proclaims it magnificent beyond poet or painter. Mingling with the terrible there is the sublime. It is a hushed Wagner, a tragedy of silent music, a Godtune terrific. Here you are face to face with the eternal.... And sentinel over all is the mountain that took its name from one who stood not upon its wind-swept crest.*[28]

Fatty Rice's Emporium
Photo, Pioneer Museum

Fatty Rice, who operated a roadside tavern outside the Gateway entrance, is reported to have called the Garden "The best sparkin' spot in El Paso County." An unidentified lady from Boston stated, "After the first few moments of wild exclamation, one sinks into awed silence." Thomas E. Townsend, an early resident, expressed his feelings on his first visit. "The effect was one of peace rather than agitation, a reverence, a peace, such as you find in old cathedrals and churches abroad.... The grand sentinels of red granite and other wonderful formations alone stood there silent in all their glory." Helen Hunt Jackson wrote in her book *Bits of Travel at Home*, "I doubt if one ever loved the Garden of the Gods at first sight. One must learn it like a new language; even if one has known nature's tongues well, he will be a helpless foreigner here. I have fancied that its speech was to the speech of ordinary nature what the Romany is among the dialects of the civilized; fierce, wild, free, defiantly tender; and I believe no son of the Romany fold has ever lived among the world's people without drooping and pining."[29]

In *The Heart of the Continent*, Fitz Hugh Ludlow describes the rocks in the Garden of the Gods:

Helen Hunt Jackson
Photo, Colorado College

This fanciful name is due to the curious forms assumed by red and white sedimentary strata which have been upheaved to a perfect perpendicular on a narrow plain at the base of the foot-hills, with summits worn by the action of wind and weather into their present statuesque appearance. There is not much garden to justify the title; but it would not be difficult to imagine some of the curious rock-masses petrified gods of the old Scandinavian mythology. These masses, upon their east and west faces, are nearly tabular. Some of them reach a height of four hundred feet, with the proportions of a flat grave-stone. Two of the loftier ones make a fine portal to the gateway of the garden. Their red is more intense than that of any of the sandstones I am acquainted with, in a bright sun seeming almost like carnelian.

Gateway Rocks,
Manitou Springs
Heritage Center,
Heine

Ludlow goes on to describe the various formations in the Garden:

One of the ... red rocks resembles a statue of Liberty standing by her escutcheon, with the usual Phrygian cap on her head. Still another is surmounted by two figures which it requires very little poetry, at the proper distance from them, to imagine a dolphin and an eagle aspecting each other across a field gules. The spine-cracking curve of the dolphin, and his nice, impossibly fluted mouth would have delighted any of the old bronze-workers.... The eagle, too, was quite striking.... Another rock resembles a pilgrim (poetical, not Plains' variety) pressing forward with a staff in his hand; another is supposed to look exactly like a griffin. Indeed, from the right point of view one feels that a griffin must very probably look thus, though the difficulty of comparing it with an original specimen prevents absolute certainty.[30]

Other visitors were moved to write poetry about the Garden of the Gods.

Balanced Rock, Garden of the Gods
J. L. McDowell 1905

Long, long ere time's relentless task began
Of measuring life by a mortal span,
Ere sun and moon, with radiance bright,
To heavenly hosts revealed the might
Of Eternal Mind in His wise plan
To create a world - and then a man -
Voiceless was the barren earth, and cold
The waters deep which o'er it rolled.

Dread glaciers ground their sullen way
And volcanic fires empoisoned day;
Then restless seas affrighted fled,
And mountains grand appeared instead.
While rugged peaks soon towered on high
To appall the sense and please the eye.

All this we know, for oft we see
Full many a witness, mute, like thee.
Oh wondrous rock on thy narrow base.
Poised, as it were, for a leap in space.
While far below winds the work of man,
He of creation's inspired plan -
Who in wonder, awe, does gaze on thee.
The work of Nature's God as well as he.[31]

Dolphin and Eagle,
Manitou Springs
Heritage Center,
Jackson

Manitou
Edgar P. Vangessen (1910)

SPIRIT OF THE STREAM:	THE SPIRIT OF THE SPRINGS AND STREAM
Sister spirit of the spring.	White tepees crown our hills,
Fresher, clearer voices sing	Sweeter lips now touch our rills;
Of a white, later race	Under Manitou's bright skies
Taking the swart Indian's place	Fairer faces meet our eyes;
Art to Nature gives her hand;	And where crystal waters glide
Fashion waves her magic wand	Happy lovers blush and hide;
And the languorous glamour cast	Dusky features fade away,
Veils the glory of the past.	Saxon faces crown To-Day.[32]

Pushing Wagon up Ute Pass
Photo, Manitou Springs Heritage Center

These descriptions as well as the discovery of gold in 1859 brought more and more people to the Pikes Peak area. A greater population required better means of travel than foot and horseback. A road was built following the old Ute Trail. In 1859, Horace Tabor, at that time a small-time grocer from Colorado City, hauled supplies up Ute Pass to the gold camps in South Park. The trip took him two weeks at five miles a day with a wagon load of flour, sugar, pork, and other trade goods.[33]

In 1925, Irving Howbert wrote that in 1860 the Ute Pass wagon road was located as it was in 1925, except that just below the present town of Cascade it left Fountain Creek and ran over the hills from a quarter to a half mile south of the creek, closely following the old Ute Indian trail down to the upper end of the present town of Manitou.[34] The wagon road from the Soda Springs to Colorado City followed the old Indian trail most of the way. This trail had been widened where it ran through thickets of chokecherry, currant, and other bushes, and was a fairly good road.[35]

On February 26, 1861, the Colorado Enabling Act was passed, making Colorado a Territory. William

Gilpin of Missouri was appointed first Territorial Governor of Colorado. Governor Gilpin appointed M. S. Beach, Henry S. Clark, and A.D. Sprague, Commissioners of the new County of El Paso and charged them with the duty of organizing a county government. "The County of El Paso, as originally laid out, was about forty by sixty miles in extent, and comprised the territory around Pike's Peak - west of that mountain almost as far as the Platte River and east for approximately thirty miles on the plains. Through the center of it, east and west - north of Pike's Peak - runs the famous Ute Pass, from which came the Spanish name of the county, El Paso."[36]

Horace Austin Warner Tabor Photo, Denver Public LIbrary

The first Territorial Legislature of Colorado met in Denver on September 9, 1861 and voted to locate the Territorial Capital at Colorado City. The second Legislature of Colorado met at Colorado City on July 7, 1862. The town had only one hotel, and it had a limited capacity, so many of the members had to camp out during their stay. They were dissatisfied, and after convening, changed the capital to Denver. The members then quickly departed and journeyed to the new capital. An amusing account of this event was given by Judge Wilbur F. Stone in an address at a meeting of the El Paso County Pioneer Society held September 25, 1906. Judge Stone was a member of the House of Representatives from Fremont County in the Legislature and took an active part in its proceedings. His account in part was as follows:

> *The first Territorial Legislature of Colorado, which consisted altogether of but twenty-two members... located the capital of the Territory at Colorado City. This was done chiefly on the ground that it was the geographical center of Colorado and the gateway to the mountains. There was also a poetical idea associated with the Garden of the Gods. Perhaps, too, some may have entertained a prosaic notion that Pike's Peak as a landmark would serve to guide the wayward members from the remotest camp to that valley of wisdom. So there in July 1862, the lawmakers met for the second session..., but by a misunderstanding as to the time of meeting, a number were not present, there being only seven or eight of the Council in attendance.*
>
> *A more unique gathering together of a legislative assem-*

bly probably never before presented a subject for chronicle. They came in wagons, horseback and on foot. Two old overland stage coaches brought loads from Denver and Gregory diggings. George Crocker and his fellow member from California Gulch footed it all the way over the snowy range, across the South Park and down through the mountains, one hundred and forty miles, carrying each a blanket on which they slept beside the trail wherever night overtook them.... I shall never forget the aspect of that brilliant man and lawyer, George Crocker, when he walked up and laid down his blanket at the door of the House of Representatives.... His dress was a blue flannel shirt, trousers patched with buckskin, an old boot on one foot and a brogan on the other, an old slouch hat that he had slept in, the brim partly gone.... The next day we elected this same George Crocker speaker of the House....

The Council assembled in the kitchen of old Mother Maggard's log tavern. The body took a recess while the cooking was in progress.... The members of the House batched in their little assembly room, took turns at cooking at the fireplace, ... carried water in pails from the creek, and slept at night on the floor.

At the end of two or three days the charm of this sort of life began to wane.... There was not enough of paper, ink, pens or postage stamps in the capital for a day's business, no printing press nearer than Denver. So the House passed a joint resolution to adjourn to Denver. This was opposed in the Council by two or three southern members, the leader of whom was the Honorable Bob Willis of El Paso County, who held a large number of shares of stock in the Colorado City Town Company, and this move threatened a fall in capital stock. The opponents of the move were in a minority, but it required their presence to make a quorum. ...before a final vote was taken, [they] fled from the capital and secreted themselves at Willis' ranch, a dozen miles down the Fountain. At the end of a day or two...two members were sent down to the ranch with a flag of truce, and proposed a treaty whereby, in consideration of a certain number of shares of capital stock, they agreed to vote against the resolution, if the absentees would return and allow the vote to be taken.... No sooner was a quorum counted than the doors were locked, the question put to vote and carried, ayes five, nays three. The three were soon laughed into submission, and in less than twenty minutes teams were hitched up, mules saddled, and the gay capital was moving from the mouth of Camp Creek to the mouth of Cherry Creek....[37]

The Pikes Peak gold rush brought more people, new trails appeared, buffalo and wildlife were killed or driven away. New towns sprung up at the base of the mountains. This was land on which both Ute and Arapaho traditionally hunted and clashed with each other, a land that belonged to the Indians by treaty. None of the tribes made strong

Possible First Territorial Capital Building
Photo, Manitou Springs Heritage Center

objections to settlement in the marginal land because they thought it might provide a buffer or screen between them and their enemies. In fact, the Arapaho left their wives and families in the protection of Denver and raided the Ute.[38] In 1850, a large band of Cheyenne and Arapaho had an all-day fight with the Ute in and around Monument Park.[39]

Many of the Indians were not too happy about the white men taking over their hunting grounds, killing the buffalo, and chasing off the other game. In 1862 the eastern frontier of El Paso County faced territory occupied by the Cheyenne and Arapaho, "no more crafty and blood-thirsty savages upon the American Continent."[40] It would have been easy for these warriors to reach the Colorado City area without being observed. However, they had a healthy respect for the fighting ability of the Ute. They knew they ran the risk of having to fight with the Ute as well as the whites if they came here. Irving Howbert felt the principal reason why the people of El Paso County were not exterminated was the close proximity of the Ute.

In the winter of 1862, Irving Howbert's family bought and moved onto a ranch on Camp Creek just above Colorado City. They also acquired the remainder of the valley up to what is now known as

Irving Howbert, 1890
Photo, Colorado College

Ute Chief Ouray, 1880
Photo, Pikes Peak
Library District,
Poley

Glen Eyrie. All of the region around the Garden of the Gods and west of the valley of Camp Creek was vacant land, and they kept their stock there during the next year or two. Howbert stated that at one time he thought he would preempt the Garden of the Gods and even complied with the preliminaries necessary to establish a legal claim, but later forfeited his right, thinking the tract of little value.[41]

In 1864, some cattle herders came into Colorado City late one evening and told of having seen near Templeton's Gap half a dozen mounted Indians who were acting suspiciously. Early the next morning an armed party went to the point where the Indians had been seen, found their trail, and followed it. It was discovered that at some time during the previous night, the Indians had been on the hill that overlooks Colorado City on the north, evidently for the purpose of determining the strength and preparedness of the settlement, with the view of a future attack. Their trail from that point led into the mountains west of the Garden of the Gods.[42]

The next day while Howbert was in town, he heard the alarming news, so he was keeping a sharp lookout for the Indians all afternoon. Shortly after sundown, his brothers were driving their stock in from the neighborhood of the Garden of the Gods. While Howbert was helping them drive the cattle into a corral near his house, he looked up the valley of Camp Creek and he saw about a half a mile away, "six mounted Indians leading an extra horse. They were going easterly along an old Indian trail that ran across the hills just south of the Garden of the Gods and out into the valley of Camp Creek through a gap in the ridge at what was afterwards known as the Chambers Ranch, then across Camp Creek and over the Mesa, crossing Monument Creek just above the present town of Roswell."[43]

Howbert rode to Colorado City to get help, then they went after the Indians. They captured the Indians just east of Monument

Creek. The Indians escaped, but some were later killed. It turned out that they were a scouting party for a planned raid by the Cheyenne, Arapaho, and other hostile Indians. Howbert and his men had prevented an attack on Colorado City.[44]

During the winter of 1865-66, three hundred Ute Indians camped for several months on the south side of Fountain Creek opposite Colorado City, and again in the winter of 1866-67, a thousand or more of the same tribe camped for several months below the boiling springs at Manitou, between Balanced Rock and Fountain Creek. Snow was deep and game was scarce and the Indians were starving. Finally their chiefs, Ouray and Colorow, made a demand on citizens of Colorado City for twenty sacks of flour and intimated that unless it was produced forthwith, they would have to march into town and take it by force. The citizens gave flour. This was the only time in the early period that Colorado City suffered from the presence of the Ute.[45]

Ute Chief Colorow
Photo, Pikes Peak Library District, Poley

More difficulties occurred with the Cheyenne and Arapaho in 1868. Cheyenne and Arapaho with letters attesting them to be peaceable Indians attacked and killed a number of settlers on Bijou and Kiowa Creeks east of Colorado Springs. Some of them penetrated into South Park by way of Ute Pass and attacked their old enemies the Ute, killing several of them. In the meantime, others had gained entrance into Colorado City by their letters, the citizens believing them to be peaceful. After their skirmish with the Ute, the supposedly friendly Indians stole all the livestock they could in Colorado City and escaped. A short time afterward, several attacks were made along Monument Creek in which a number of white settlers lost their lives.[46] After the battle of the Washita in November 1868, most of the Cheyenne and Arapaho were put on a reservation. This ended the raids in this area, and the people of El Paso County no longer worried about attacks by Indians, although spasmodic outbreaks of Indian violence occurred elsewhere in the west for several years.[47]

The Ute continued to frequent the Garden of the Gods area through the 1870s. Their summer encampment of 1878 was the last because of the treaty made in Washington DC with Chief Ouray in which the Ute agreed to go to reservations. One summer more than 500 Ute camped in a cottonwood grove along Camp Creek. While staying in the Garden of the Gods, the Ute would have horse races on the narrow, level plain on the mesa between Colorado Springs and the Garden of the Gods. The track, several miles in length, was worn into two tracks or ditches in the hard surface about four feet wide and two feet deep. Braves were also observed running down and catching antelope.[48]

With worries about Indian attacks minimized, the people of Colorado City could go on about their lives. The Garden of the Gods went through several changes of ownership. Although Beach and Cable had original claims along Camp Creek, no one actually

Bottom Row: Ute Chief Ignacio, Carl Shurz, Chief Ouray, Chipeta, Top Row: Woretsiz, Gen. Charles Adams
Photo, Pikes Peak Library DIstrict, Poley

claimed the Garden of the Gods until Bill Garvin, who ran a store for Garrish and Cobb in Colorado City. He appears to have been the original claimant. He laid the foundation for his dwelling immediately west of Gateway rocks as near the center of his claim as possible.[49] Fitz Hugh Ludlow in *The Heart of the Continent* states: "At one time Mr. Garvin had set his stake in the Garden of the Gods, intending to enjoy the luxury of ownership in that great natural curiosity; but other business prevented his carrying out his plan of a large house there, and, not to interfere with actual settlers who might wish the spot, he finally withdrew his claim."[50] Beach and Cable also abandoned their claims. However, because the Garden had no water, it did not appeal to incoming settlers; they had to be practical and locate where they had sufficient water to raise crops. Others attempted ownership, but usually for grazing, quarrying of gypsum, or similar purposes.[51]

Beach's claim was later taken by George Bute, then by Walter Galloway who worked in town but built a log cabin on his quarter section. In 1874, Robert Chambers bought the Galloway homestead for $1400. He built a new house of stone, put in six acres of asparagus, and planted hundreds of apple and cherry trees. He dug a reservoir on the backside of Niobrara Ridge for water irrigation. Traces of the fruit trees and the reservoir can still be seen today.[52]

Lewis N. Tappan
Photo, Denver Public Library

Like the Garden of the Gods, the springs at Manitou underwent several changes of ownership. Before Colorado City was founded, Dick Wooton and friends had claimed the boiling springs and built a log cabin to prove their proprietorship.[53] According to "Professor" Goldrick in 1859, "[Wooton] intends to erect improvements for accommodations of those that will before long seek this spot for the purpose of repairing their health with the aid of its mineral waters." Wooton sold the Soda Springs in 1859 to Whitsitt and Company

Gen. William Jackson Palmer
Photo, Colorado College

for $500. Whitsitt sold his interest to one of the Tappan brothers from Boston. The acreage was jumped by a man named Slaughter who built a house of some pretensions. Slaughter lost his rights to Thomas Girten, owner of Sulfur Springs in South Park, who obtained legal rights by jumping Slaughter. Girten was in the mineral springs business and made improvements in the Soda Springs. Girten sold out to Colonel Chivington (of Sand Creek fame) for $1,500. Chivington sold it for $500 to his son-in-law Pollock. Pollock sold the springs eventually to a Denver man, George Crater, who paid $10,000.[54]

One visitor who came to the area in 1869 was General William Jackson Palmer. In a letter he wrote in 1869 as he was traveling north from Pueblo to Denver he stated, "At Colorado City, the Garden of the Gods, we stopped to breakfast. I freshened up by a preliminary bath in the waters of the Fountain. Near here are the finest soda springs and the most enticing scenery. I am sure there will be a famous summer resort here soon after the railroad reaches Denver."[55]

When General Palmer arrived, the price of Soda Springs was up to $26,000. Palmer's Colorado Springs Company bought the springs. The Colorado Springs Company founded the resort town of LaFont, which was later changed to Manitou,[56] a name inspired by William Blackmore. Blackmore was interested in Indian lore and had just read Longfellow's *Hiawatha*. He saw Ute Indians at the

Manitou
Photo, Manitou Springs Heritage Center

Soda Springs, decided they used it for religious purposes, and that it should be named Manitou for the Great Spirit Manitou (the Algonquin Indian spirit of Longfellow's epic).[57] Buildings were built, the springs were encased in iron pipe, and water was piped directly into bottles which were distributed and sold in Denver and Pueblo.[58]

General Palmer became interested in a possible route for his railroad, so he returned to Colorado City in July 1869. In a letter dated Denver, July 28, 1869, he wrote, "Dr. Bell and Colonel Greenwood were along and the following day after spending some hours at and around the springs, we rode thru the Garden of the Gods into Glen Eyrie and across by Monument Park to Teachout's where we spent the night, and then rode on to Denver, sleeping, I remember, in a haystack one night on West Plum Creek."[59]

On yet another visit in October, General Palmer wrote, "On Cherry Creek, October 21st [1869]. Yesterday we came hither from Colorado City. The whole party was in the finest humor. Pikes Peak never looked grander, and the Garden of the Gods fascinated my companions of the Eastern frontier [Mr. N.C. Meeker, Cyrus Field, James Archer] so that they bubbled all over with enthusiasm, resembling the Soda Springs, from which we drank great quaffs, as Dr. Bell and I had done only a month before."[60]

During the spring of 1870, on a tour prior to selecting the site for Colorado Springs, General Palmer,

Mineral Analysis Chart, Manitou Springs Heritage Center, Anna Galbreath

Dr. William C. Bell,
Photo, Colorado College

Colonel Greenwood, Palmer's future wife, Queen Mellen, and others were visiting the area. Irving Howbert accompanied the party on horseback to the Soda Springs, the Garden of the Gods, and other places of interest in the Pikes Peak region. Palmer was trying to convince his fiancée that she could be happy living in the Pikes Peak Region.[61]

In 1871, Palmer hired William E. Pabor to publicize the attractions around his new development of Colorado Springs, planned as a resort city on the Denver and Rio Grande Railroad. Pabor, Secretary of General Palmer's Colorado Springs Company, in his effort to attract people to the area, described the formations in the Garden in exotic terms. According to Earl Pomeroy in *In Search of the Golden West*:

> *The rock formations rather significantly known as the Garden of the Gods, near Colorado Springs, had a typical and dependable appeal, despite the misgivings of some old-fashioned spirits who felt that "something less heathenish (for a name) would better have befitted these Christian days." In some respects they represented the extremity of interest in the grotescque. "No other point of interest," boasted the Denver and Rio Grande Railroad, "is more unique or more to be admired than this curious freak of nature." Its admirers thought that they saw (or at least said that they saw) a stag's head, "curious birds and crawling serpents," "an eagle with pinions spread," a seal making love to a nun, and an elephant attacking a lion, among other "grotesque and picturesque sights." "There is little doubt," complained one visitor, "that the average tourist is so intent on finding these monstrosities, that he miss the grandeur and glory of the place." More significantly, perhaps, the tourist saw the Old World as well. He might "pass under the shadow of China's great wall, muse among Palmyra's shattered and fallen columns, stand face to face with the mysterious Sphynx of Egypt, gaze upon the Temples of Greece, or the Castles of England and Germany, or the old Abbeys which pious monks upreared." The exchange of Western curiosity for European antiquity had become almost literal.*[62]

Queen Palmer
Photo, Colorado College

Even though tourists flocked to see "Seal Making Love to a Nun," Palmer was not amused and asked Pabor to resign.[63]

Seal and Bear
Photo, Manitou Springs Heritage Center

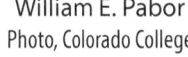

William E. Pabor
Photo, Colorado College

Below Left: The Hobgoblin
Below Right: The Simpleton
Photos, Manitou Springs Heritage Center

Above: Cathedral Spires
Photo, Old Colorado City Historical Society

Below Left: Mother Grunde
Below Right: Griffin
Photos, Colorado College

At the same time General Palmer was concerned with the image of the Garden of the Gods, scientists were making an important discovery beneath the surface of the Garden of the Gods. James Huchison Kerr, an 1865 Yale graduate moved to Colorado Springs in 1874, seeking a cure for his tuberculosis. In 1875, a healthy Mr. Kerr became the professor of assaying, geology, mining, metallurgy, and chemistry at Colorado College and served as acting president from 1885 to 1889.[64] In 1878, Professor Kerr "discovered in one of the ridges east of the red rocks forming the east boundary of the Garden of the Gods portions of 21 different sea monsters that had been caught in a basin in one of Earth's early paroxysms."[65] One of the animals whose remains were found is estimated to be 117 feet long and of the cretaceous period.[66] The bones were stored in boxes placed in barns and cellars around the college. Nearly all were subsequently lost.[67] Professor Kerr contacted Dr. O. C. Marsh, a noted Yale paleontologist about the find. Dr. Marsh and a crew excavated what he believed to be a Camptosaurus skull and took it back to Yale Peabody Museum where it was stored away and forgotten for 110 years.[68] The excitement over the scientific discovery was brief, but visitors continued to be enthralled by the beauty of the Garden of the Gods.

James Huchison Kerr
Photo, Colorado College

In 1877, Isabella Bird, Author of *A Lady's Life in the Rocky Mountains,* rode through Glen Eyrie and the Garden of the Gods and wrote, "I then came to strange gorges with wonderful upright rocks of all shapes and colours, and turning through a gate of rock, I came upon what I knew must be Glen Eyrie, as wild and romantic a glen as imagination ever pictured. The track then passed down a valley close under some ghastly peaks, wild, cold, awe-inspiring scenery."[69] She goes on to say that the names of the places near the Great Gorge of Manitou are familiar to everyone, the Garden of the Gods, Glen Eyrie, Pikes Peak, Monument Park, and the Ute Pass.[70]

General Palmer's Residence
Photo, Pikes Peak Library District

General Palmer had no interest in ownership of the Garden of the Gods because it had no water. Instead, he chose Glen Eyrie, so named because of the eagles' nest on the great gray entrance rock. Palmer named the canyon to the west, Queen's Canyon, after his wife.[71]

On a journey from Denver to Colorado City, Albert D. Richardson and his traveling companions passed unusual red rock formations on the way south, culminating in the towering Gateway to the Garden of the Gods. They spoke of eagles building their nests on the summit and of the great cave in the Gateway Rock. Richardson wrote of the experience:

"Through this natural gateway we passed into a large enclosure walled in by mountains on every side—indeed a garden for the gods. One vast rock has a cave eight feet by sixty and about seventy in height. Its walls are smooth and seamless.

"We entered by the only aperture, barely large enough for an adult to crawl through. Within we struck a light to view the weird picture. For an hour the singers of our party made the walls echo with the strains of sacred music, always most impressive in underground chambers."[72]

Not all visitors were favorably impressed with the Garden of the Gods at first sight. Ernest Ingersoll, writing of the sights of Ute Pass in his travel journal *The Crest of the Continent:* said, "Then there is the... Garden of the Gods, hidden behind those garish walls of red and yellow sandstone, so stark and out of place in the soberly-toned landscape that they travesty nature, converting the whole picture into a theatrical scene, and a highly spectacular one at that. Passing behind these sensational walls, one is not surprised to find a sort of gigantic peep-show in Pantomime..."[73] Isabella Bird and her hosts paid a visit to General Palmer at Glen Eyrie:

After lunch, the -s in a buggy and I on Birdie left Colorado Springs, crossing the Mesa, a high hill with a table top, with a view of extraordinary laminated rocks, leaves of rock a bright vermilion colour, against a background of snowy mountains, surmounted by Pike's Peak. Then we plunged into cavernous Glen Eyrie, with its fantastic needles of colored rock, and were entertained at Genera Palmer's "baronial mansion," a perfect eyrie, the fine hall filled with buffalo, elk, and deer heads, skins of wild animals, stuffed birds, bear robes, and numerous Indian and other weapons and trophies. Then through a gate of huge red rocks, we passed into the valley, called fantastically, Garden of the Gods, in which, were I a divinity, I certainly would not choose to dwell. Many places in this neighborhood are also vulgarized by grotesque names.[74]

General Palmer
on Horseback
Photo, Colorado College

Julian Street wrote in his book *Abroad at Home* of his 1919 visit to the Garden of the Gods:

The place irritated me. For if ever any spot was outrageously over named, it is that one. As a little park in the Catskills it might be all well enough, but as a natural wonder in the Rocky Mountains, with Pike's Peak hanging overhead, it is a pale pink joke. If I had my way I should take its wonder-name away from it, for the name is too fine to waste, and a thousand spots in Colorado are more worthy of it.

The entrance to the place, between two tall, rose-colored sandstone rocks may, perhaps, be called imposing; the rest of it might better be described as imposition. Guides will take you through, and they will do their utmost, as guides always do, to make you imagine that you are really seeing something. They will point out inane formations in the sandstone rock, and will attempt to make you see that these are "pictures." They will show you the Kissing Camels, the Bear and Seal, the Buffalo, the Bride and Groom, the Preacher, the Scotsman, Punch and Judy, the Washerwoman, and other rock forms, sculptured by Nature into shapes more or less suggesting the various objects mentioned. But what if they do? To look at such accidentals is a pastime about as intelligent as looking for pictures in the moon, or in the patterns of the paper on your wall. As nearly as Nature can be altogether silly she has been silly here, and I think that only silly people will succeed in finding fascination in the place—the more so since Colorado Springs is a prohibition town.[75]

General Palmer loved the red rocks, and after settling at Glen Eyrie, he kept urging his friend and fellow railroad builder Charles E. Perkins to build a home in the Garden of the Gods and to bring his railroad, the Burlington, from Chicago to Colorado Springs.

In 1879, Perkins bought from J. H. Hayes the 240 acres which surround the main Gateway Rocks, including Garvin's original claim, eighty acres bordering to the south, and 160 acres to the west, totaling 480 acres for $10,473. He never built the home due to the scarcity of water and his lack of time, and the Burlington never reached Colorado Springs. For twenty-five years, Perkins left the management of the property to Henry Wills. For twenty-five years, he paid the taxes, oversaw the removal of trash and debris, and kept the Garden of the Gods open and free for all visitors.[76] However, this purchase was not enough to protect the attraction from various entrepreneurs.

In 1879, a gypsum quarry was located on the back side of White Rock. A trench-like scar is still visible today. In the late 1870's, a lager saloon was built under the image of the kissing camels, and in March 1883, Billy Bryan built a stairway to the top of Gateway Rock. By July, he had opened a nearby resort and celebrated the Fourth with fireworks and a moonlight dance. Transportation to the Garden of the Gods was limited, so some entrepreneurs and

Charles Perkins
Photo, Colorado College

speculators developed plans to remedy the situation. In 1889, The El Paso Rapid Transit Company had been incorporated. They obtained franchises to run horse cars to the Garden of the Gods as well as to Austin Bluffs, Colorado City, Manitou, Cheyenne Canyon, and other points of interest. In 1892, Fatty Rice, a colorful character who weighed 300 pounds, acquired four corner lots in the newly-platted Garden City development a block or so east of the Gateway Rocks, and in 1898, acquired two more corner lots. Here Fatty established an emporium where he sold liquid refreshments "of the strong type," food, and curios to all who came by, whether by horse, carriage, hiking, or cycling up over the Mesa following the little white markers that General Palmer provided to enable him to find his way back to Glen Eyrie at dusk when riding his horse home from the office in Colorado Springs. Palmer and Perkins felt that Fatty's place, even though outside the garden proper, distracted attention from the "dignity and beauty of the surroundings."[77]

Garden of the Gods, Sightseeing Trip
Photo, Pioneer Museum

Refreshment Stand
Photo, Pikes Peak Library District

Fatty Rice had competition for his curios in the form of Charles Strausenback. In 1900, 10-year-old Charles E. Strausenback set up a tri-pod table near the Gateway Rocks to sell his gypsum carvings and painted rocks.[78] Charles was born in 1890 in Morelia, Michoacán, Mexico of a German father and American mother.[79] After several moves, the family settled in Manitou Town (Manitou Springs), Colorado. By the time he was 19, the tripod had become a booth to sell curios including rock slabs painted with Garden

Charles Strausenback selling gypsum carvings in 1900
Photo courtesy, Garden of the Gods Trading Post

of the Gods images and Western designs. After high school, he worked as a buyer in Arizona for the Fred Harvey Co. which dealt in American Indian art and artifacts. In 1914, Charles returned to the Pikes Peak area and resumed selling his curios and artwork. For a time, he worked for Curt Goerke taking photographs at Balanced Rock and managing some of Goerke's businesses located on the boundary of the Garden of the Gods Park.[80] In 1919, Charles married Esther Isabell Brumbaugh who also became his business partner.[81]

In 1929, Charles and Esther built a Pueblo themed trading post on Becker's Lane at the park boundary.[82] They advertised that "Real Indian curios will be carried, including Indian silver,

Navajo blankets, pottery, etc."[83] One of the main attractions in the curio shop was a "petrified" Indian, (a rock statue wearing a breechcloth) which was on display until 1979. Other attractions included Pueblo and Navajo craftsmen whom Charles hired to demonstrate silversmithing, jewelry making, rug weaving, and

Trading Post c.1929
Photo courtesy, Garden of the Gods Trading Post

pot making. The Indians lived in their dwellings at the Trading Post, giving tourists a glimpse of "normal" Indian life. Awa Tsireh (Afonso Roybal) of the San Ildefonso Pueblo, Ca-Ping (Severo Tafoya) of Santa Clara Pueblo, and William Goodluck, a Navajo, worked as silversmiths at the Trading Post seasonally until about 1955.[84] During the winters from 1936 to 1948, the Strausenbacks lived in Phoenix, Arizona where they bought merchandise for the Trading Post and ran the Hotel Adams Gift Shop which they renamed Strausenback Indian Silver Shops in 1940.[85]

Charles Strausenback briefly studied art with Broadman Robinson who introduced him to the Modernist movement. Robinson went on to become the Director of the Colorado Springs Fine Art Center. Charles was a very versatile artist, showing influences from his background in Mexico and southwestern United States, often using pseudonyms to sign his work. Charles Ernest Strausenback was his real name and Charley Earnesta (Modernist), Charley Yazza (Navajo style art), and Ontya (Pueblo style art) were his pseudonyms.[86]

Charley Earnesta painting by Strausenback,
Photo courtesy, Garden of the Gods Trading Post

In October 1957, Charles Strausenback had a stroke and died. Esther continued to run the Trading Post until 1979. At that time, she sold the business to the Haas family's TAT Enterprises which continues to run the Trading Post today. The store, originally 2,000 square feet, was expanded to 25,000 square feet, containing a general store, restaurant, art gallery, gold panning, and arcade games while staying true to the original building and its nearly century of history. The walls of the Trading Post are decorated with Native American art as well as paintings and photographs by Charles who also designed the logo on the front of the building.[87]

Trading Post logo
Image courtesy, Garden of the Gods Trading Post

Charles E. Strausenback
Photo courtesy, Garden of the Gods Trading Post

Orchard House
Pikes Peak Library District, Standley

Rockledge Ranch
Pikes Peak Library DIstrict

In 1907, Palmer, who had earlier purchased many of the Garden City lots as a buffer zone for the Garden,[88] including his 1900 purchase of the Chambers Ranch, purchased the property from Fatty's widow. On the Chambers' property, Palmer built the Orchard House, later called the White House, for his wife's half-sister Mrs. William Schlater. After the General's death in 1909, the property changed hands several times before finally being incorporated into the city park system as White House Ranch Historic Site, now known as The Rock Ledge Ranch Historic Site. This site consists of the 1868 Galloway house, the 1895 Chambers Rock Ledge Farm, and the 1907 Orchard house.[89]

The biggest plan was announced on July 31, 1895. A group of New York and Philadelphia men had organized a strong company to build a two-mile-long streetcar line from Prospect Street and Colorado Avenue in Colorado City to the Garden of the Gods to cost around $100,000 and to be known as The Garden and Glen Electric Road. These same speculators also planned to build an immense Palm Palace in front of the Gateway, constructed largely of glass, somewhat like The Horticulture Building at the Worlds Fair in which were to be permanently maintained exhibits and specimens of every known plant from all over the world. The Palm Palace was to cost nearly $500,000. To this there was to be added a casino, a restaurant, and so forth.[90] These ideas never materialized due to the high estimated cost.[91]

A citizen wrote a letter of protest in reaction to this plan:

> *It is scarcely conceivable that the Board will for a moment listen with favor to such a request, but experience teaches that the best opposition to such schemes is that which makes itself known early in the contest. The subject is entirely too obvious to be overlooked.*
>
> *The bare thought of such violation of the natural beauties and wonders of the Garden of the Gods ought to arouse immediate and earnest protestation from citizens generally. For years the Garden has been one of the far-famed attractions of the vicinity and it cannot be denied that much of this distinction has been due to its careful preservation in a natural state. The introduction of such a line as is proposed, with its surely attendant evils, would be a calamity inflicted without the slightest recompense. A drive heretofore rendered pleasant by its very freedom from the noise of electric travel, would be destroyed. The Garden being already within easy reach by carriages, which can be secured at very moderate rates, not one good argument can be advanced for the construction of this line.*[92]

On March 8, 1886, the Honorable Frank C. Bunnell, of Pennsylvania, introduced a bill (H.R. 6580) in the first session of the Forty-ninth Congress entitled "To preserve and protect the Garden of the Gods, in Colorado from spoilation, and to create a public park of the same." Yellowstone Park had been created in 1872 and was the only national park at this time. The only national memorials were in Washington D.C.- Washington Monument (Jan 1848) and the Lincoln Museum (April 1866). Garden of the Gods National Park as proposed was to be thirty miles square, to include Pikes Peak, Colorado Springs, Colorado City, Manitou Springs, Cripple Creek, and Victor. Colorado Springs had been founded fifteen years before. The bill was read two times without debate, referred to committee on Public Lands, but was not reported out. A comment made in the newspaper was to the effect that since the Garden of the Gods was private property, it seemed beyond the power of the Government to acquire it.[93] It is not surprising that the Garden of the Gods received such national attention when one considers what had been written about it.

General Palmer kept an eye on the Garden for Perkins who had left it open to the public. Palmer stopped a scheme to extend the streetcar line along his bridle path up Camp Creek to the Garden. On April 3, 1903, Palmer received a letter from Robert McReynolds of Colorado City suggesting it would be a good idea to carve busts of some the Presidents in the giant peaks in the Garden of the Gods. He thought it would make it an everlasting mecca for travelers from all over the world. It would be another wonder of the world. The writer thought it would be such a drawing card that railroads would contribute large sums of money to bring it about. He suggested the busts be of Lincoln, Garfield, and McKinley. General Palmer replied that he felt it would not be in good taste to mutilate the scenery. (Mt. Rushmore was completed in 1927.) He wrote to Perkins telling him of the suggestion.

Perkins, in response, wondered why Mr. McReynolds left out Roosevelt.[94]

On October 27, 1906, a few weeks after General Palmer's seventieth birthday, the General and his daughters Dorothy and Marjory, and a friend had been horseback riding in the Garden of the Gods. They were headed home when just a mile or two from Glen Eyrie, the General's horse stepped on a rock and stumbled. Palmer fell and landed on his head on the road. The fall had broken his neck in three places and injured his spinal cord. He could only move his head and neck freely, move his shoulders, and bend his elbows slightly. Palmer lived three more years and then passed away in his home at Glen Eyrie at the age of 72.[95] General Palmer and Charles Perkins' dream of preserving the Garden of the Gods for the free use of everyone has lived on, and millions of people from all over the world have benefited from their generosity.

Endnotes:
1. "Lost Cave in Garden of Gods Found and Sealed Up; Earliest of Pioneers Carved Names and Dates There," *Sunday Gazette and Telegraph,* (Colorado Springs, Colorado: Sunday, October 13, 1935), 4.
2. Gehling, 10.
3. *Ibid.,* 7.
4. Cragin Papers, XVIII, (Colorado Springs Pioneers' Museum), 10.
5. Garden of the Gods Folder, (Colorado Springs, Pioneers' Museum).
6. *Ibid.*
7. M. S. Beach, "El Paso County, the First Settlement." Colorado Springs Pioneers' Museum.
8. *Ibid.*
9. Garden of the Gods Folder.
10. *Ibid.*
11. *Ibid.*
12. *Ibid.*
13. Clara E. Belschner, "Pioneer Tells of Ox-Cart Trip to C.S. Back in 1865" (*Sunday Gazetter and Telegraph,* November 3, 1935), 1-2.
14. Garden of the Gods Folder.
15. *Ibid.*
16. *Ibid.*
17. Gehling, 8.
18. Woodard, 6.
19. Stone, 146.
20. Woodard, 6.
21. Gehling, 11.
22. Woodard, 6.
23. *Ibid.,* 6.
24. *Ibid.,* 6.
25. Howbert, *Memories,* 83.
26. Woodard, 6.
27. J. E Liller, *El Paso County* (University Place, Nebraska: Charles Elce and Son, 1924), 7-9.

28. *Ibid.*, 7.
29. Helen Maria Fiske Hunt Jackson, *Bits of Travel at Home* (Boston: Roberts Brothers, 1878), 217-218.
30. Woodard, 8.
31. Earl Spencer Pomeroy, *In Search of the Golden West; the tourist in western America* (New York: Alfred A. Knopf, 1957), i.
32. *Ibid.*, i.
33. Fitz Hugh Ludlow, *The Heart of the Continent: a record of travel across the plains and in Oregon, with an examination of the Mormon principle* (New York: Hurd and Houghton, 1870), 178-180.
34. Howbert, *Memories*, 37.
35. *Ibid*, 43.
36. *Ibid.*, 69-70.
37. *Ibid.*, 73-75.
38. J. Donald Hughs, 56-57.
39. Howbert, *Memories*, 213-214.
40. *Ibid.*, 213-214.
41. *Ibid.*, 83.
42. *Ibid.*, 101.
43. *Ibid.*, 102.
44. Howbert, *Indians, 78-92.*
45. *Ibid.*, 72-73.
46. Stone, 99.
47. Howbert, *Memories, 214.*
48. Charles Cross, *Tourist Guide and Poetic Description of the Garden of the Gods, Glen Eyrie, Manitou and Vicinity* (Colorado Springs, 1906), 9.
49. Gehling, 11.
50. Ludlow, 183.
51. Woodard, 39.
52. Gehling, 12.
53. Liller, 4.
54. Manly Dayton Ormes, *The Book of Colorado Springs* (Colorado Springs, The Dentan Printing Co., 1933), 23.
55. *Ibid.*, 23.
56. Fetler, 144.
57. Sprague, 34-35.
58. Fetler, 144.
59. Ormes, 29.
60. *Ibid.*, 29.
61. Howbert, *Memories,* 219-220.
62. Pomeroy, 54-55.
63. Howbert, *Memories, 47-48.*
64. Finding Aid to the James Hutchison Kerr Papers MS 0081, Colorado Springs: The Colorado College Tutt Library, 1949, 2.
65. *Ibid.,* 4.
66. "Professor Kerr Finds Fossils –*Scientific American*." *Colorado Springs Gazette,* 23 Feb 1878. p 4.

67. Finding Aid, 4.
68. Melissa Walker, "Dwelling Among the Wonders and Mysteries of Life." *The Bulletin: Summer '98:* (Colorado College Publications), 1998.
69. Isabella L. Bird, *A Lady's Life in the Rocky* Mountains (New York: E. P., Dutton & Co., 7th edition, 188?), 177.
70. *Ibid., 180-181*
71. Sprague, 97.
72. Albert D. Richardson, *Beyond the Mississippi: From the Great River to the Great Ocean; Life and Adventure on the Prairies, Mountains, and Pacific Coast, 1857-1867* (Hartford: American Publishing Co., 1867), 312.
73. Ingersoll, 46-47.
74. Bird, 182.
75. Julian Street, *Abroad at Home: American ramblings, observations and adventures* (Garden City, New York: Garden City Publishing Co., 1914), 427-428.
76. Gehling, 15.
77. Woodard, 10.
78. Messier, Kim and Pat, Images of America, Garden of the Gods Trading Post. Charleston; Arcadia Publishing, p. 18.
79. 1900 Twelfth US Census, Topeka, Kansas.
80. Colorado County Marriage Records and State Index, 1862-2006.
81. Trading Post. Square Space.com/history. Accessed 7/24/23.
82. "New Indian Store in Garden of the Gods.", Colorado Springs Gazette, April 7, 1929.
83. Messier, Pat and Kim 2014 Reassessing Hallmarks of Native Southwest Jewelry. Schiffer Publishing pp 41-45.
84. Kim & Pat Messier's Blog, "Strausenback as Artist. Part II: Lithographs and Etchings," May 30, 2022."
85. "Strausenback Wins Renown as Artist." Transcription from Colorado Springs Gazette, Jan 8, 1936, p. 12.
86. Kim & Pat Messier's Blog," Strausenbacks as Artist, Part I: Original Art, January 2, 2023.
87. trading-post.squarespace.com/history.
88. Gehling, 12.
89. *Ibid., 12.*
90. Woodard, 10.
91. Gehling, 16.
92. Woodard, 10-12.
93. *Ibid.*, 20.
94. C. E. Perkins, "Dear Sir," (William Jackson Palmer Papers, Colorado Historical Society, Denver, Co.) The *Colorado Prospector, Vol. 17, No. 4)*, 3.
95. Sprague, 155-156.

The Garden of the Gods as a City Park

Charles Perkins died in Massachusetts in November 1907, leaving the Garden to his two sons and four daughters. Shortly before his death, Perkins reportedly scribbled on an old envelope his desire to give the Garden of the Gods to the city of Colorado Springs. Two years later, his six children turned the Garden of the Gods over to three trustees until the legal transfer could be completed. A few days before 1909, the Colorado Springs City Council voted to accept the following restrictions that accompanied the gift:
1. The park is to be forever known as the Garden of the Gods.
2. No intoxicating liquors can be sold or manufactured there.
3. No buildings will be erected there, except those necessary to properly maintain the area as a public park; it will be forever free to the public.

Any violation of these restrictions will result in the property ownership reverting to the Perkins heirs.[1]

An article in the December 25, 1909 issue of the *Gazette* announced the gift in the following manner:

> *When Colorado Springs awakes this morning, it will find in its stocking the biggest Christmas gift in the history of the city - the famed Garden of the Gods, presented by the late Charles E. Perkins. This marvelous park, which is known throughout the United States and to many foreign lands for its curious rock formations, today becomes the property of Colorado Springs.... The property, comprising 480 acres, is valued at $300,000 and makes the city's park system the most varied in the world, for a community this size. The ordinance under which the gift was accepted by Colorado Springs bars the sale and manufacture of any intoxicating liquor on the property. The erection of buildings except those deemed absolutely necessary for park purposes, is also prohibited.*

A formal ceremony of acceptance was held on October 3, 1912.[2] The accepting ordinance was as follows:

> ...in accepting the deed to this marvel of nature the City of Colorado Springs, through its Mayor and Council, hereby expresses its deep, sincere and abiding sense of gratitude to the donors, Charles E. Perkins, Robert F. Perkins, Alice Perkins Hooper, Edith Perkins Cunningham, Margaret Perkins Rice, Mary Russell Perkins and all those who have united in presenting this great gift to the City, and hereby assures them that the City of Colorado Springs, and all the inhabitants of, and visitors to the Pike's Peak Region shall always hold each of them in kind and loving memory for this gift of inestimable value and the City of Colorado Springs joins with those who will have their hearts made glad by the Garden of the Gods in the future, in sincerely wishing that the choicest and most satisfying blessings of life shall ever be the inheritance of each of the generous donors and his posterity and their posterity to the latest generation.[3]

Plaque Dedication
Photo, Pioneer Museum

A tablet was unveiled at the Gateway entrance to the Garden of the Gods, called the "Beautiful Gate" by the pioneers. Mrs. Charles Elliott Perkins was on hand for the occasion. She later wrote to her children, "This morning... I drove through the Garden of the Gods, and I recalled the time when your father and I confidently expected to pass our summers there... What dreams we had, your father and I, and how we planned for the future. How young we were and how confidently we planned, plans that I am trying to carry out alone... But unless one is absolutely devoid of sentiment how can one be here in this lovely spot, read the Tablet, and not dream again?"

Of the ceremony itself, she wrote:

Plaque Honoring
Charles Elliott Perkins
Photo, Colorado College

The occasion was most impressive, but I cannot do justice to the ceremony. At the time of the unveiling, Rob was on the stand with the Park Commissioners, Judge Lunt, Dr. Donaldson, the Mayor, and Ex-Mayor Hall. Old Sol stood close to me, on the running board of our automobile, and we mingled our tears. Robert Forbes Perkins made a short address, and pulled the string, and the tablet was unveiled. We all stood up, and Sol said out loud, "That's nice." The sun was in my eyes and gave me an excuse for holding up my sunshade! It was pretty trying, most touching, and I was grateful when it was over. I thought of your father every minute... Much feeling has been shown, and I wish all you children could have been at Colorado Springs, for you would then understand what your gift means to the friends there.[4]

The Garden of the Gods was such a popular place that certain citizens felt better transportation was needed into and around the park. In the *Colorado Springs Gazette* of April 25, 1913 came the announcement that "...resolutions advocating the establishment of a permanent summer carnival ground in the Garden of the Gods, and recommending the construction of an electric line to the eastern terminus of the park, were adopted last night by the Colorado

Above: Paul Goerke
Below: Curt Goerke
Photos, Pioneer Mueum

Goerke Photography Business, November 16, 1926, Photo, Old Colorado City Historical Society, Goerke

Springs Boosters Association." None of these propositions came to pass. Automobiles made it easier to reach the Garden - its isolation was fast disappearing.[5]

The 480 acres given by the Perkins family was increased in later purchases by the City of Colorado Springs until the park reached today's size of 1364 acres.[6] Among the earliest acquisitions was Mushroom Park, including Balanced Rock and Steamboat Rock. In January 1923, Colorado Springs paid Curt Goerke $25,000 for 275 acres which he and his father had developed on the western edge of the Garden of the Gods. For over thirty-five years, Goerke and Son had carried on a very popular photography business centered around Balanced Rock. After the Kodak camera came into popular use, young Goerke built a high board fence around Balanced Rock and charged twenty-five cents admission.[7] The early photos were of people standing beside little burros in front of the rock (on the south side).[8]

The Garden of the Gods has long been a favorite place for picnics, games, and festivals. Day-time picnics for young people included games like one called "snake scramble." Young ladies scrambled through the scrub oak and when they saw a rattlesnake, they had to scramble back to their starting place. The last girl in had to kiss somebody.[9]

The 1910's brought an economic slump to Colorado Springs. To offset this slump, some citizens sought to draw tourist trade by staging a historical pageant about Indians and the sixteenth century ac-

tivities of the Conquistadors in New Mexico and of old King Tartarax of Quivera. This pageant culminated in an amateur rodeo at the Garden of the Gods.[10] The first celebration was held in 1911 and was marked by a parade of two hundred floats decorated with flowers, bands, and five volunteer fire departments; a masked ball; an aviation event featuring a bi-plane flown in for the event; and a group of Ute Indians from the Ute Reservation in Ignacio, Colorado were invited. Chipeta, wife of the famous chief Ouray, and Chief Buckskin Charley, two notable leaders of the Ute Tribe, accompanied the group.[11]

Ladies on the Rocks
Photo, Colorado College

Rodeo Contestants, 1912
Photo, Pikes Peak Library District

Amateur Rodeo,
Photo, Old Colorado City
Historical Society

In 1912, the Ute were brought back from the Reservation to the Pikes Peak Region to dance in the Garden of the Gods and place offerings in the springs at Manitou, as their ancestors

Left: Flower Carnival Entry
Photo, Pikes Peak LIbrary District

Below: Chipeta in Carriage,
Photo, Manitou Springs Heritage Center

Chipeta, Wife of Chief Ouray
Photo, Pikes Peak Library District, Poley

Chief Buckskin Charlie
Photo, Pikes Peak LIbrary
DIstrict, Poley

had for generations. This gathering enabled visitors unfamiliar with Indian life and customs to have the opportunity to see the Indians living under what was advertised as normal tribal conditions.[12] The Ute Indians were the major attraction, and the celebration was named Shan Kive (Shawn Keedie), supposedly meaning heap good time in Ute.[13] A platform was built high up on South Gateway Rock about fifty or sixty feet above the ground so that all could see the Indians perform their traditional dances. Seventy-five Indians came into view in single file up over the top of a steep, narrow, winding trail which began at the extreme southern end of the Rock. Accompanied by tom-toms and their own singing, the men and women put on their famous Sun Dance, followed by their Harvest Dance and others.[14] The main event of the 1912 celebration was the marking of the Ute Trail. The Indians were driven to Cascade where they mounted horses, and accompanied by Irving Howbert, David N. Heizer, and E. E. Nichols, all prominent citizens, they rode down over the ancient Indian trail in the hills on the south slope of Ute Pass. "The procession started down the trail from Cascade toward Manitou, a colorful procession. The braves in their buckskin suits and feathered war bonnets, and the women greatly adorned with turquoise and silver jewelry. The Indians needed no guides. All the Indians were gay and happy. They chatted in the Ute language and were singing their

Ute Dance
Photo, Pikes Peak Library District, Poley

Ute Chiefs at Shan Kive, 1913
Photo, Pikes Peak LIbrary District, Stewarts

Left: Shan Kive Crowd
Pikes Peak Library District,
Stewarts

Below: Ute on the Move
Photo, Pikes Peak Library District,
Powell

Above: Ute on Old Ute Trail

Below: Ute Trail Marker
at Manitou Springs
Photos, Pikes Peak Library District,
Poley

songs. Buckskin Charlie, chief of this tribe, had not been over the trail since the Ute had left this country over thirty-years earlier. His birth place was the Garden of the Gods and he remembered every turn in this trail he had ridden since he was a boy. 'I'm seventy years old,' he said, 'I never so happy in all my life.'"[15]

Marble markers were placed along the route (vandals destroyed the marble markers, so concrete markers replaced them). When the group reached Manitou, ceremonies were held at Soda Springs Park where dignitaries spoke to several thousand in the audience. A box containing historical mementos was placed under a boulder. (This box, in a much rusted condition, was opened in 1962 to reveal a soggy disintegrated pulp.) In 1913 the celebration boosters wanted a permanent site in the Garden of the Gods; however, the Indian camp was located near Adams Crossing (between Manitou and Colorado City, where the red rocks opened into a natural amphitheater on private property). At the camp, a wild-west midway was created featuring one hundred Indians and five hundred cowboys. Buckskin Charlie and Chipeta were again the guests

Ute Celebration
Photo, Old Colorado City Historical Society

Ute in the Garden
Photo, Pikes Peak Library District, Poley

of honor. On the first day was a ten-mile footrace from the park through city streets, then the participants climbed Mesa Road to end at the Garden of the Gods. Eighteen runners started and twelve finished with the winning time of seventy minutes. A grand parade was featured on the second day when 30,000 spectators watched as fifty Ute, three hundred cowboys, seven hundred pioneers, and Civil War Veterans marched through downtown. The final two days had bucking bronco contests, foot races among the Indians, Ute dances in the Garden, and a carnival in the streets. The festivities ended with a masquerade ball.[16] The 1913 celebration was the last visit by the Ute. The trip to Colorado Springs was too expensive and against Indian Agency rules because the event commercialized the Indians. The carnival continued without the Ute for more than a decade, until the start of World War I. After 1920, the celebration evolved into the Pikes Peak or Bust Rodeo.[17]

The National Woman's Party held their convention in the Garden of the Gods on September 23, 1923. Susan B. Anthony, Elizabeth Cady Stanton, and Lucretia Mott took part in a pageant at Pulpit Rock before what was reportedly the greatest crowd ever assembled in the Garden of the Gods. The pageant consisted of women in hoop skirts, pantelettes, and poke bonnets, as well as twenty-four young girls on horseback representing the youth of the Western world, which was to be the vanguard of the drive for the Equal Rights Amendment to the federal constitution drafted by Alice Paul. An enthusiastic appeal for support of Western women in the forthcoming congressional campaign was made by Mrs. Oliver H. P. Belkmont, president of the National Woman's Party.[18]

National Women's Party Pageant
Photo, Pikes Peak Library District

During the early twentieth century, the cavern in North Gateway Rock had been forgotten and the mouth became filled with weeds and debris until 1935 when an old man who had played there as a child insisted upon the reopening of the cave. At first, the parks people did not believe him, but he prevailed and the cave was found and reopened. The name A. S. Vorhees as well as those of several other early pioneers was still visible. The Civilian Conservation Corps carted out 175 loads of dirt and cut steps in the cavern floor. However, falling rock spoiled the plans to open the cave to the public. The cavern was permanently sealed with concrete and iron bars. (It was again opened briefly in 1953.)[19] In 1935, the CCC also planted several thousand junipers and obliterated a number of dirt roads that marred the beauty of the Garden of the Gods.[20]

According to the agreement with the Perkins family at the time the Garden of the Gods was given to the city, only a few buildings were permitted in the park. The first one built in the Garden of the Gods had special significance to the people who visited it. Located west of North Gateway Rock and nestled between two fifty-foot sandstone wings was the Hidden Inn, a pueblo style building designed by Carl Balcomb and constructed by Colorado Springs architect Thomas P. Barber. Built in 1915, the Hidden Inn was designed as a curio shop and tea room offering souvenirs, food, and drink, but no intoxicating liquors as mandated by the agreement with the Perkins family. R.

Civilian Conservation Corps Worker
Photo, Pikes Peak Library District

Cathedral Canon with Young Junipers, 1935
Photo, Old Colorado City Historical Society

Hidden Inn, 1920s
Photo, Manitou Springs Heritage Center

S. Davis held the concession to operate the lunchroom and the curio shop where Charles Strausenback sold his carvings and artwork. The Inn also served as living quarters for the caretaker of the park. Visitors to the park could find refuge here during storms, browse through the souvenirs, enjoy a snack, or look out at the sky and red rocks from the roof-top terraces. This building was a fine example of early pueblo architecture, being based on the early Indian adobe houses and the Taos Indian Pueblo.[21] The Hidden Inn and the Trading Post were the only structures not demolished when the property was taken over by the City Parks Department in 1909.[22]

In 1948, after Davis had retired, Helen Wilson Stewart with her company H. W. Stewart Inc. took over operation of the Hidden Inn.[23] Helen's family, the Wilsons, had moved to Colorado from Kansas in the early 1890's. In 1892 they began publishing the "highest newspaper in the world," the *Pikes Peak Semi-Daily News*, on Pikes Peak at the end of the newly completed Manitou and Pikes Peak Cog Railway. Helen was a true entrepreneur, and as a child, she would sell fresh-picked mountain flowers and baked goods to the Cog train passengers, and then she would board the train for the ride down the mountain to sell new editions of the paper to tourists in Manitou. She had trained her horse Brownie to follow her and the cog train down the mountain so she could ride him back home. As a young woman, Helen became a Western Union telegraph operator on the summit of Pikes Peak in what became known as the World's Highest Telegraph Office.

Helen's future husband Orrie Stewart worked for J. G. Heistand, the "Official Photographer of the Manitou and Pikes Peak Railway" and had a photography business on the summit. Orrie and his brother Ben both graduated from Colorado College where Orrie also served as a chemistry professor from 1909-1910. Taking advantage of Orrie's chemistry background, Orrie and Ben started "Stewart Brothers Photo Finishing and Burro Rental" on Ruxton Ave in Manitou. Helen and Orrie married October 15, 1919, and had three daughters. By 1935, the Stewarts had received a special permit from the forestry service to operate the Pikes Peak Summit House and the Halfway House at Glen Cove. These two operations included lunchrooms and the sale of curios and postcards. Hotel rooms and a gas station were also available at Glen Cove. The lunchroom on the summit sold the famous Pikes Peak doughnuts made from Helen's grandmother's recipe.

Orrie passed away in 1939 and left Helen the task of running the Pikes Peak operation. During the Great Depression, Helen had to sell the family home and buy an apartment house to provide consistent income for her family. During World War II, times were tough due in part to the rationing of gasoline, and Helen had to ask for help. Charlie Tutt, a prominent businessman, made a commitment to the U.S. Army at Camp Carson that the Cog Railway would continue to operate so the soldiers would have something to do during their time off. He promised to cover any losses Helen might incur. After World War II, Helen's daughter Barbara and her husband William F. "Bill" Carle became active in the business and the post-war prosperity enabled them to expand it to new heights. The Stewart family obtained leasing and licensing agreements for the operation of concessions throughout the state of Colorado, including Pikes Peak Summit, Glen Cove Inn, and the Garden of the Gods' Hidden Inn.

In the 1992 the Stewart family lost the Concession on Pikes Peak because larger national corporations were allowed to bid, and Aramark was awarded the contract. The Hidden Inn was demolished by the city of Colorado Springs in 1999 in accordance with the Garden of the Gods Master Plan.[24]

The popularity of the Garden of the Gods extends beyond recreation. The Garden of the Gods holds a special spiritual connection for many of the people who have encountered it. The Reverend Albert W. Luce of the Pike's Peak Christian Church used to ride his bicycle into the Garden of the Gods for inspiration for his sermons. One day, a few weeks before Easter, he was sitting at the foot of Cathedral Spires reading his Bible.

The words of John 19:41-41 had special meaning: "Now in the place where he was crucified there was a garden: and in the garden a new sepulcher, wherein was never man yet laid. There they laid Jesus..." The thought came to him that this was a garden just like that mentioned in the Bible, and where would it be more appropriate to hold the Easter service than in that which so nearly resembled the one at Gethsemane, as he imagined it? Reverend Luce next thought about a time for the service. Again the Bible, John 20:1 told him. "The first day of the week cometh Mary Magdaline early when it was yet dark, unto the sepulcher..." Rev. Luce decided sunrise would be the most appropriate time to hold the service. Just before Easter in 1919, he announced his plans to his congregation. They were stunned. He went ahead with the service and was joined by about seven hundred people, most from his church, some from the First Christian Church. This service became very popular. Soon it was too much work for him, so he made it a community affair, turning it over to the Colorado Springs Ministerial Alliance. The service became an interdenominational service to which all were invited, with around 20,000 to 25,000 attending (1950), and was broadcast coast to coast over radio (Columbia Broadcasting System,

Easter Sunrise Service
Photo, Old Colorado City Historical Society

Easter 1930s
Photo, Colorado College

The Armed Forces Network, and The Voice of America).[25] Fear of damage to the fragile ecology inside the park, the service was moved to the east side of the Gateway Rocks.[26] By 1990, the numbers began dwindling and the nation-wide coverage was suspended.[27] After seventy years, the last service was held in 2002 to a crowd of 5,000.[28]

Chuckwagon Dinner
Photo, Pikes Peak Library District, Stewarts

Beginning in 1935, the Colorado Springs Jaycee's and their wives began hosting family-oriented chuck wagon dinners in the Garden of the Gods. The first was held for a group of eight visitors to the region. Over the next three years, the dinners were held in different places in the park. Within those three years, the popularity had grown to the point where the dinners needed a permanent home and the sponsors needed to plan on a larger scale. The area just east of the North Gateway rock was chosen and facilities were built to handle a few hundred people. Dinners were given Mondays, Wednesdays, and Fridays from mid-June through Labor Day. Seating had to be limited to seven hundred to keep the dinners informal and so everyone could see the activities. Serving was completed in twenty-five to thirty minutes. The menu was hot biscuits, butter, vegetables, meat, potatoes, salad, coffee, and dessert. Everyone was seated at a table with a view of the campfire and small stage off to one side. The campfire would be lit in the dramatic Flaming Arrow Ceremony. A flaming arrow was "shot" off the top of one of the

Robert Mitchum
Photo, Pikes Peak Library District, Stewarts

Gateway Rocks to light the campfire. Following dinner, wranglers would bring in a hollering and struggling victim for the branding scene. This "victim" was usually a Jaycee, or sometimes a visiting celebrity. A white-hot branding iron was applied to the seat of his jeans (he had a board inside his pants), and he would go running out through the crowd with a red-hot JC on his seat. Following the branding was "rough-shod, rip-roaring, and delightful" entertainment put on by the wranglers in true Western style with lots of jokes and group singing.[29] These dinners were an institution in the Garden of the Gods until the late 1970's. The pavilion was then used for reserved shows and as a showcase for "Legend and Legacy," a one-hour show by the Colorado Springs Park and Recreation Department, portraying the early days of the Pikes Peak region.[30]

During the early days, students from Colorado College went dancing and picnicking in the moonlight in the Garden of the Gods.[31] Other uses of the Garden of the Gods have included civic gatherings, foot races, rock climbing exhibitions, and weddings.

Jane Russell
Photo, Pikes Peak Library District, Stewarts

Horseback riders, hikers, and mountain bikers have also found the Garden of the Gods a special place to practice their sports. In the 1940's, the Garden was a favorite stop-over for recruits on their way to war, and in the 1960's, the Garden became a favorite haunt for the Flower Children.[32]

In August of 1968, while flying over Brazil, General Charles A. Lindburg wrote a letter to the Garden of the Gods Preservation Council expressing his strong conviction that the surroundings of the unique Garden of the Gods should be protected from further man-made encroachments. With his letter he sent an unsolicited gift of $100 to aid in preservation of this natural land. In the letter, the General reminisces about his visits to, and flights over, the Garden of the Gods:

The Garden of the Gods means a great deal to me, for it, and

Colorado, have woven through my family's life, and mine for many years. One of my great uncles on my mother's side took a major part in the protection of the fish in Colorado streams. I feel sure he admired the Garden of the Gods as I do in a later generation.

My own first contact with these fantastic rocks came in 1916, when I was driving my mother, and another uncle, on a trip between Minnesota and California. I was then fourteen years of age.

My mother had been telling me about the great Rocky Mountains, and the Garden of the Gods, even before we left our Minnesota home. I remember watching the colors unfold as we approached the area - slowly, because the roads then were unpaved and rough. When we arrived at the rocks themselves, we stopped for several hours, and I climbed one of them - my little Fox-Terrier, "Wahgoosh," following me. I remember a very narrow-ledge trail in one spot, with a precipitous fall at one side. I think I have never seen a more spectacular and magnificent place.

Barnstorming in Colorado, in 1925, I flew over and circled the Garden of the Gods; and laying out the transcontinental passenger airline in 1928-29, I often detoured far enough from my route to fly above it. I think my last airview of the Garden (at least from reasonable low altitude) was from the plane carrying the Air Force Academy site-selection commission, of which I was a member. But my clearest memories came from earlier years, when there was enough time to climb or circle.

The Garden of the Gods is one of the most beautiful and spectacular areas in the world. It would be a tragedy of major magnitude to lose, or to detract from, its great qualities - an inexcusable indifference on the part of our generation, and a lack of responsibility for those succeeding us. I sincerely hope the citizens of Colorado are successful in their efforts to preserve it.[33]

Soldier's View from
the Hidden Inn
Photo, Manitou Springs
Heritage Center

Others were of the same mind as General Lindbergh. In December 1968, Colorado Springs officials received a surprise gift of

Trail Signs
Author's Collection, Evans

$300,000 from two local foundations to purchase a forty-two-acre tract adjoining the Garden of the Gods to provide a "buffer zone" for the city's famed tourist attraction. The El Pomar Foundation gave the city $250,000 and the Bemis-Taylor Foundation gave $50,000 for the land purchase. Both foundations, in accompanying letters, stressed the need for the immediate purchase of the tract. El Pomar's letter called for the purchase by December 31. The forty-two acres, a part of Vrooman Estates including what is now Rockledge Ranch Historic Site, had been approved for a multifamily development on approximately twenty-eight acres of the tract. The development was to consist of eighty-three one-story townhouse units, sixty-four two-story apartments, and two three-story multi-family dwellings with 107 units. The Garden of the Gods Preservation Council were overwhelmed by the gift, and, as a consequence, redoubled their efforts to raise the two million dollars to purchase the remaining 485 acres of privately owned land encircling the Garden of the Gods.[34]

In February 1969, Stuart W. Richter, director of the City Park and Recreation Department and the Park and Recreation Advisory

Rock Climbing
Author's Collection, Evans

Board recommended that the White House Ranch (Rockledge Ranch) and the property which borders the Garden of the Gods in the recent purchase, be used as an Ecology Center. Its future would be as a working farm-ranch model so that groups and individuals visiting the area could appreciate the fact that residents of the Pikes Peak Region at one time raised farm produce on this property. The early owners had supplied the Antlers Hotel with fruits, vegetables, and preserves. Richter was against turning this property into a golf course, saying the area should be preserved as much as possible in its natural state and used as a center for the study of the natural environment with overtones of tying in the historical past since the Ute Trail ran through the area and that the home itself was built by General William J.

Palmer, the city's founder.[35]

The local American Indian population continues to use the Garden of the Gods for religious purposes. On June 21, 1997, the second annual World Peace and Prayer Day at the Garden of the Gods was sponsored by an American Indian group called N.A.T.I.V.E.S.. Arvol Looking Horse, nineteenth generation keeper of the White Buffalo Calf Pipe set this day aside for all Indian Nations to come together for the healing of Mother Earth. This was a spiritual gathering and all Indian Nations were invited to attend. Spiritual leaders, political activists, and authors attended, and those who were participating danced, sang, and held a spiritual honoring of the ancestors and Mother Earth.

Balanced Rock and Bikers
Author's Collection, Evans

Students and their teachers are another group who find the Garden of the Gods a special place. A group of earth science students from Horace Mann Jr. High School were browsing through the park one day in 1959 when two boys found a message hidden in a crevice under a rock a short way above Balanced Rock. The message was from a man named Charles Walscavage, who had walked to the Garden of the Gods from Pennsylvania in 1924. According to the note, Walscavage had crammed it into the crevice more than thirty-five years earlier.[36]

Dinosaur Fossil, Theiophytalia Kerri, Found in the Garden of the Gods
Author's Collection, Evans

In 1994, a group of paleontologists, geologists, and artists under the direction of Melissa Walker, then program coordinator for the Garden of the Gods, collaborated on research for the murals in the visitors' center. During the course of this research, Dr. Ken Carpenter, vertebrate paleontologist for the Denver Museum of Natural History, sent Ms. Walker information about a dinosaur fossil that was discovered in the Garden of the Gods in 1878. This fossil turned out to be Professor Kerr's forgotten dinosaur. A call to the Yale Peabody Museum Collections Manager confirmed that the Colorado Springs Camptosaurus fossil skull was still there.   In late1996, Dr. Kirk Johnson, one of the

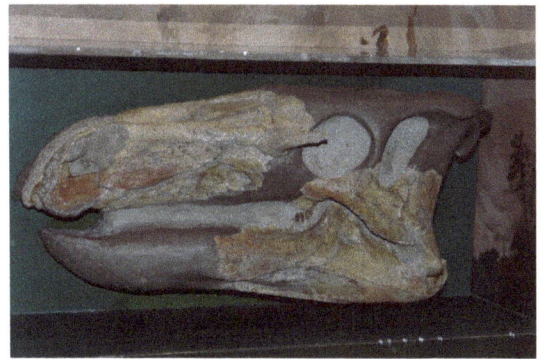

advising paleontologists hand carried the irreplaceable skull back to Denver. The staff made a cast of the original skull and presented it to the City of Colorado Springs on behalf of the Denver Museum of Natural History. The replica was placed on exhibit at the Garden of the Gods Visitors' Center and a Camptosaurus was depicted in the mural of the Jurassic Age.[37]

While the display at the visitor center was being prepared, Dr. Carpenter and his colleague Kathleen Brill started a new investigation of the 100-million-year-old bones. They discovered the soil adhering to the skull was actually from the Cretaceous period. Upon closer examination of the skull, they noted several differences between this skull and a Camptosaurus. Ten years after first seeing the fossil, Carpenter and Brill were confident enough in their research to name the creature Theiophytalia kerri and declare a new genus and species, named after the Garden of the Gods and Professor Kerr. The cast of the skull is now on display at the Garden of the Gods visitor center. Drs. Carpenter and Brill continue to research this new species in Dinosaur Ridge outside of Denver in hopes that it may help clear up the mystery of unidentified footprints from an unknown iguanodon. Because of the damage it would cause to the park, the paleontologists have no plans to dig for more bones in the Garden of the Gods.[38]

Because some of the popular uses of the Garden of the Gods are damaging to the environment or dangerous to the participants, they have had to be regulated. The rocks of the Garden of the Gods are soft and crumbly, and inexperienced rock climbers, and occasionally experienced as well, find themselves in difficult situations. Police, fire, and park officials have been rescuing people off the rocks almost since Colorado City was first founded. Climbing deaths are rare at the park, but they do happen occasionally. Rock climbing is allowed at the park, but climbers must sign in at the visitors' center and pledge that they have the necessary safety equipment. Those types of climbers usually have little problem and do not attract the scrutiny of park safety patrollers. It is the tourist, the young kids, and the over-confident college students who have the most trouble.[39]

In the 1970's horseback riding was regulated, prohibiting the activity from the garden zone just south of the old Hidden Inn. Riders now have a network of trails encircling the Garden Zone. Mountain biking has become increasingly popular in the last decade, and because the bikers are fond of bushwhacking (going off the trail), much of the Garden has been damaged. These eroded trails are being blocked off with pine boughs in an attempt to protect the area, and now there are rules regulating off-road biking.[40] Bikers too have a designated area in which to practice their sport in the beauty of the Garden of the Gods without endangering the environment.

The Garden of the Gods attracts thousands of visitors each year, and the negative impact on the environment was sufficient to bring together a group of concerned citizens to form a master plan for preserving the beauty of the park.

Endnotes:

1. Gehling, 15.
2. Sprague, 141.
3. Woodard, 39.
4. *Ibid.*, 39-40.
5. Woodard, 12.
6. *Ibid.*, 41.
7. Gehling, 20.
8. Woodard, 37.
9. *Ibid.*, 99.
10. Sprague, 254-255.
11. Woodard, 18.
12. *Ibid.*, 18.
13. Virginia McConnell, "Real Indians Highlighted Shan Kive: Festival Predecessor to the Rodeo" (*The Free Press*. Colorado Springs, Colorado: late 1960's), 4-5.
14. Woodard, 18.
15. Pettit, 26.
16. Gehling, 18.
17. McConnell, 4-5.
18. "Act Out Fight For Suffrage," *The Rocky Mountain News* (*Colorado Prospector: Colorado History From Early Day Newspapers, Vol.* 17, No. 4), 6
19. Gehling, 10.
20. "New Roads in Garden of the Gods?" *Gazette and Telegraph* (Colorado Springs: Sunday, November 3, 1935), 1.
21. Michael H. Collins, "Historic Structures Report for the Hidden Inn, Garden of the Gods" (For the City of Colorado Springs Parks and Recreation Department, 1996)
22. Gehling, 21.
23. "Helen W. Stewart: Colorado's High-Altitude Hostess 1893-1979" *West Word*, Old Colorado City Historical Society. Vol 30, No. 6 August 2015.
24. "Colorado tourism's first family…," Jason Blevins, jblevins@denverpost.com, May 29, 2016.)
25. Woodard, 30-31.
26. David Hughes, "The Pastor's Bicycle and His Bible." Colorado Springs.yourhub.com, 8/22/2008.
27. Gehling, 18-19.
28. David Hughes.
29. Woodard, 33-35.
30. Gehling, 24.
31. Sprague, 281.
32. *Ibid.*, 19.

33. General Charles A. Lindbergh, Letter to the Garden of the Gods Preservation Council, August 24, 1968 (Garden of the Gods Folder, Colorado Springs Pioneers Museum).
34. Jim Gibney, "Generous Christmas Gift," (*Denver Post*) *Colorado Prospector: Colorado History From Early Day Newspapers* (Vol. 17, No. 4), 11.
35. "Ecology Center Proposed," *The Colorado Springs Gazette and Telegraph* (*Colorado Prospector:* Vol. 17, No. 4), 11.
36. Ted Wilcox, "Garden of Gods Keeps Message 37 Years," *Gazette Telegraph* (Colorado Springs: 1961).
37. Melissa Walker, "Dwelling," 3.
38. Bill Reed, p 10.
39. Todd Hartman, "Climbers ignore stone-cold fact," *Gazette Telegraph* (Colorado Springs: June 18, 1997), 1A.
40. Barry Noreen, "Mountain biking ban to be enforced," *Gazette Telegraph* (Colorado Springs: July 3, 1993).

The Master Plan

Attracting over a million visitors a year, the Garden of the Gods has begun to show the wear and tear of so much use. A Master Plan for preservation of the Garden was completed in 1976; however, by 1993 some of the issues of concern in 1976 were still unresolved, and new issues were also being added to the list. The issues of traffic, rock-climbing, concessions, visitor center, erosion, and trails still were not resolved, and the new issues of mountain bikes, signage, and accessibility needed to be addressed. The Parks and Recreation Department, which is responsible for caring for the Garden, established a planning process to update the Master Plan of the park.[1]

The planning process for the Garden of the Gods Master Plan Update began in February 1993. The people at the Parks Department decided to make the update process an open or fishbowl process with no designated committee. This meant that all concerned citizens and staff would work together to develop the plan. The initial meeting was held on June 9, 1993 at the Chuckwagon Pavilion in the Garden of the Gods. A mailing list of 125 grew to one of 450 citizens, groups, or participants who were involved in developing the new Master Plan. A special meeting was also held to discuss Native American use of the Garden of the Gods Park.[2]

The initial meeting informed the participants of the need for the plan. The technical aspects of the plan followed the Design with Nature process developed by Ian McHarg. This process evaluated the environmental and historical aspects of the park: vegetation, wildlife, aesthetics, geology, archeology, history, soil, hydrography and gradients. Other studies included traffic, user and community surveys, and user conflicts such as auto-pedestrian or auto-bicycle.[3]

The guiding principle framing all discussions about the future was "Conservation, preservation, and restoration are overriding principles, and, within those principles, the ultimate aim is to allow activities in the park which will not conflict and which are appropriate to the setting."[4] Citizen focus groups were created and input was gathered on issues in five major categories: Traffic, Rock Climbing, Management, Trails, and Interpretation. The recommendations of the focus groups were the primary basis of the final master plan. The issues and concerns discussed at the Native American meeting were also incorporated into the master plan. The final meeting of the master plan group was held on April 7, 1994. In all, more than 400 citizens attended fifty-one public meetings which were conducted on various aspects of the Garden of the Gods Master Plan Update in 1993 and 1994. The final Garden of the Gods Master Plan 1994 was approved by the Parks and Recreation Advisory Board on May 19, 1994. The Colorado Springs City Council voted final approval of the Garden of the Gods Master Plan on August 9, 1994.[5] One of the many issues addressed in the master plan was the need for a new visitors' center.

In September 1993, Lyda Hill of Dallas, Texas unveiled a proposal to build a new visitor center for the Garden of the Gods. The plan, which envisioned a $3.5 million complex on 30th Street east of the park, was generally supported; however, some people were wary of the project. Some of the terms of the proposal were that Hill, whose family owns the Seven Falls tourist attraction, would contribute $75,000 a year to a foundation, run by an advisory group comprised of representatives of the Parks and Recreation Advisory Board, the White House Ranch Living History Association, and the Friends of the Garden of the Gods, established to maintain the park and to pay for park guides (at that time paid for by the city). Hill also planned to operate a tram service through the park, thus relieving traffic congestion in the park. In exchange, the city would grant Hill an exclusive contract disallowing any retail activity within the park in accordance with Charles Perkins' original intent to have no commercial activities within the park boundaries. A financial analysis estimated that Hill's offer could bring $600,000 in revenue to the city during the next twenty years.[6] Because the visitors' center technically would be outside the park, (due to a land swap with Lyda Hill) this would be in accordance with Perkins' intent.

The Friends of Garden of the Gods, the League of Women Voters, and many other community activists supported Hill's proposal which received the unanimous endorsement of the Park and Recreation Advisory Board on November 18, 1993. However, opponents circulated petitions in opposition to the Hill proposal and Manitou Springs merchants who felt threatened by the thought that Hill would have the best location to sell souvenirs to tourists filed complaints.[7]

On Tuesday, December 7, 1993, the City Council unanimously approved the Hill plan for a visitors' center at Garden of the Gods. The decision was not easy, as public sentiment was far from unanimous since many Native Americans and Manitou merchants were not in favor of this plan. Hill's attorney, Bruce Warren, told the council, "For those who might say 'Bah humbug,' to this gift, I would say that this is a wonderful holiday gift to us, our children and our grandchildren."[8] According to the *Gazette Telegraph* on December 12, 1993, "Emotions run high where the Garden of the Gods is concerned. That's because some see it as a spiritual shrine, while others view it as a tourism cash cow and still others perceive it to be merely a city park."[9]

The objections to the proposed visitors' center were many and varied. One resident who lives close to the Garden of the Gods said the Hill proposal "is not a gift but a business deal," and urged the council to delay its decision until the master plan was completed. Another critic worried that Hill would unearth the remains of American Indians buried in the Garden of the Gods long ago. Another asserted that beneath the red rocks are ancient Mayan ruins. A representative of the Earth Spirit Pagans compared the Garden of the Gods to a cathedral and objected to tourists taking pictures. Although a representative of the Lone Feather Council had endorsed Hill's proposal, several individual American Indians said the local group did not speak for them.[10]

Garden of the Gods Visitors Center
Author's Collection, Evans

The proposal also had supporters at the meeting. Terry Haas, owner of Garden of the Gods Trading Post on the park's southern boundary, said Hill's plan would be good for the Garden of the Gods. He supported it 100 percent even though Hill would be in direct competition with him. Other supporters included Friends of Garden of the Gods, the Sierra Club, and the White House Ranch Living History Association. Some of the positive aspects of the plan included the guaranteed annual cash flow to maintain the park, the tram system to reduce the summertime traffic flow, and a visitors' center and parking lots located away from the rock formations. When asked why she was doing this project, Hill responded, "My 'why' is to protect the park. This is a way for me to walk my talk, a way to do something that will last."[11]

In January 1994, a group of American Indians complained they had been forgotten as the city planned the future of the Garden of the Gods. Members of the Native Hearts Council said their long history in the Garden of the Gods had been ignored when the City Council approved Lyda Hill's plan to build a new visitors' center next to the park. They were also irritated about the Master Plan effort. They asserted that "The Native American community was really unaware of what was going on."[12]

The Native Hearts Council group wanted the city to acknowledge that the park does not just occupy a niche in American Indian lore, but is sacred ground, deserving of special protection and status. They also maintain that parts of the park were once used as a burial ground. Mel Locklear, an American Indian who has lived in Colorado Springs for more than thirty years, questioned why Indian powwows involving several hundred participants were disallowed in the mid-1970's, while thousands are allowed to worship on Easter every year. The American Indian group also wanted to ban rock climbing in the park because the rocks are sacred.

Terry Putman, a Parks and Recreation Department planner, said the city could do a better job of conveying the historic role of American Indians at the park, but he said no archaeological study at the Garden of the Gods has confirmed that it was ever used as a burial site. Some members of the Indian group vowed to continue researching Smithsonian Institute records to show that Indian remains were taken from the park around the turn of the century.[13] The only documented remains were those of an ancient Indian woman which were removed from the Garden of the Gods for study in 1935 and then later reburied in the park.[14]

In spite of the protests, city officials were ready to go ahead with the visitors' center project. However, in mid-January 1994 an appeal was filed by a group called the Public Awareness Citizens Committee to block the project. The basis of the appeal was that the land swap between Lyda Hill and the city to allow for the realignment of 30th Street violated the terms of two deeds which added acreage to the park in the 1970's. Those deeds contained restrictions stating that the land must always be used for park purposes. City Attorney Jim Colvin maintained that the parking lots and road realignment were being done to accommodate the park.[15] The appeal was rejected by the City Council.

The Citizens Committee filed another appeal late Friday, February 25, 1994. The defendants named in the appeal were the City Council, the planning commission, the Park and Recreation Department, a local consulting company, and a local company formed by Lyda Hill. Attorney Mary Thrower alleged that the council "sold the farm" in an open-ended contract with Hill, and said the contract violates the city charter by allowing Hill an exclusive franchise to operate a tram service in the park without the approval of voters.[16]

The suit was filed so late on Friday that Hill's representatives were not able to appear. Thrower, in her effort to obtain the court order to stop construction without opposition, did not notify the City Attorney's Office, which was instead notified by the court. Thrower also asked that the press be barred. That request was denied. Judge Campbell said he could not stop the project on Friday because the opponents could not show that Hill's development company would not be damaged by a delay. Judge Campbell met with attorneys for all parties on Monday, February 28, and denied the request to declare the visitors' center contracts null and void.[17]

On Friday March 4, 1994, Fourth Judicial District Judge Richard Hall denied the request for an injunction to stop construction filed by Sandra Turcotte-Schaeffer, who claimed the Garden of the Gods is an Indian burial ground and sacred place subject to protection under federal law, specifically the Native American Grave and Repatriation Act. Turcotte-Schaeffer also claimed that chemicals from a golf course planned nearby by Lyda Hill would harm herbs that grow in the park and are used in American Indian ceremonies. Several witnesses testified that they use the park for ceremonies and private prayer. However, in the hearing it was shown that ceremonies occur at other areas within the park and will continue, even if the visitors' center is built. No evidence about golf course chemicals was presented.

Assistant City Attorney Stephen Hook pointed out that the federal law cited by Turcotte-Schaeffer applies only to federal lands and Indian reservations. In his judgment, Judge Hall said, "I am not persuaded that this construction will affect the ability of Native Americans to worship." Hall also said he had seen no evidence that the five acres in question had any archaeological ruins or burial grounds on it, and, in fact, only an old cow bell was discovered during the excavation and archaeological survey process. After this judgment, Thrower indicated she intended to file an amended complaint partly based on the contention that the City Council violated the City Charter by accepting a $490,000 road improvement loan from Hill without making a specific plan to pay it off. However, road grading had already begun on the project which was projected to be completed by the spring of 1995.[18]

Plans for the controversial Garden of the Gods visitors' center were formally unveiled on Thursday, April 7, 1994. The plans called for a 12,600 square-foot building scheduled for completion by May 1995. So that it would blend in with the surrounding area, the plans called for the building to be tinted red to match the park's rock formations. The visitors' center would be fronted by a 200-car parking lot. The visitors' center would take place of both the old visitors' center and the Hidden Inn, which for seventy-nine years had housed a gift shop and snack bar. About a third of the new visitors' center building was planned to hold educational displays about the park's plants, animals, geology, and American Indian history. It would also have a 3,000 square-foot gift shop, and a 3,000 square-foot dining room and kitchen. An eighty-seat theater which would show documentaries and an observation terrace were also planned.[19]

The construction of the new Garden of the Gods visitors' center left the question of what to do with the old Hidden Inn. More than fifty people showed up at a meeting on Tuesday, April 5, 1994, to discuss the future of the pueblo-style structure which had been built in 1915. Like most issues involving the park, agreement was hard to come by. Failing to reach a consensus, the group agreed to meet again for more discussion. What to do with the Hidden Inn became a point of contention within the park master planning process as well. Representatives of the Sierra Club felt the city should honor its long-term goals of reducing traffic and removing buildings from the park. Pat Conley of the Sierra Club said, "We went into this process wanting to keep it [the Garden of the Gods] natural and remove structures." The Sierra club also wanted most of the pavement near Hidden Inn torn up and the ground returned to a natural condition. Another city resident added, "People come to the Garden to see the place, not any man-made building that was put there." Several other groups wanted Hidden Inn to stay. Friends of Garden of the Gods said that Hidden Inn had been there long enough to qualify as part of the park's history. They agreed that there should be no retailing at Hidden Inn, but that using it as an education center made sense.[20]

At a meeting held in 1997 to determine the fate of the Hidden Inn, representatives of the Native American community envisioned a year-round hands-on cultural center where park visitors could learn about Indian history. Terry Putman agreed that the Hidden Inn could be used as an American Indian cultural center, with upkeep funded by Friends of Garden of the Gods, Indian organizations, and other groups. Putman said the building could have space for exhibits, restrooms, and offices for park workers. Other groups said that the Hidden Inn should house displays about the park's flora, fauna, and geology.

Alternatives for the Hidden Inn which were discussed in the meeting included: Removing the building at a cost of about $30,000; removing only the non-historic additions to Hidden Inn built after the initial construction in 1915; renovating the entire building for use as an American Indian cultural center at an unknown cost; using the building as a natural-science education center; building an all-new American Indian cultural center, regardless of the future of Hidden Inn; keeping the parking lot west of Hidden Inn regardless of the building's fate.[21]

The final outcome of the meeting was to let the Hidden Inn stand for another two years (until 1997) to allow for more study of the issues. City officials long favored removing most of the buildings from the park including the High Point building on the park's south side, the old visitors' center, and the Chuckwagon Pavilion. However, the Hidden Inn was excluded from this because of the considerable community support for saving it.[22]

As an effort to save the Hidden Inn, architect Michael Collins prepared a report on the feasibility of restoring the Hidden Inn to meet safety and handicap accessibility codes. The report states in part:

> The Hidden Inn is a unique and beautiful edifice that has been part of Garden of the Gods Park and the Colorado Springs community for over eighty years. Positioned quietly and subtly between two wings of red sandstone, the Inn is a marriage of natural and human design. The outcroppings add beauty to the building; the pueblo style copies the color and follows the form of the sandstone; the views from the upper floors of sky and stone and flora bring visitors to a deep appreciation of nature.
>
> Discussions with the Colorado Historical Society indicate that the Inn will easily qualify for the National Register of Historic Places....
>
> Michael H. Collins, Architect, through historical research, interviews with local citizens, structural, mechanical, electrical analysis, and on-site inspections has concluded, as stewards, we should rehabilitate and restore the Hidden Inn to continue its permanent use by the Colorado Springs community.[23]

Walkers
Author's Collection, Evans

In 1997, meetings again were held to decide what to do with the eighty-two-year-old Hidden Inn. The N.A.T.I.V.E.S. group at this time proposed turning the building into an American Indian Cultural Center which would contain educational exhibits on the Indian involvement in the history of the Pikes Peak area. The estimated cost of operation was $150,000 a year. On Tuesday, August 26, 1997 the Colorado Springs City Council decided the building should be torn down and re-

Joggers
Author's Collection, Evans

Bicyclist
Author's Collection, Evans

placed with natural vegetation. The cost of renovating the Hidden Inn to meet safety and handicapped-accessibility codes would cost nearly $400,000. At the meeting only one person lobbied the council, and he wanted the Inn torn down.

Eugene Red Hawk-Orner, a spokesman for the Lone Feather Council, a coalition of American Indian nations said, "This building is now the last visible offensive place. Removal of that inn would be a very strong statement of reconciliation with the various nations who revere the Garden." The N.A.T.I.V.E.S. group mounted strong opposition to Red Hawk-Orner's ideas. The feeling was also not unanimous among the non-Indian population. A citizen who works at the Eagle Stop Grocery Shop not far from the park said, "It's a pretty good old building. Even though they have the new visitors' center, I just don't think they should get rid of it." A visitor from St. Louis, who was on a bike ride through the Garden of the Gods on August 26, 1997, said, "I just hate to see what's been here a long time get torn down. Soon there won't be anything left that's been here a long time. As a tourist, that's what people want to see when they go somewhere - the historic sites - and they keep tearing them down."[24] Efforts to save the Hidden Inn based on its having qualified for the National Register of Historic Places were also blocked when the City Council and City Parks officials refused to sign the paper work to complete the process. The final plan called for the building to be removed before the tourism season started in 1998.

The removal of the Hidden Inn is part of the city's master plan for the Garden of the Gods, which includes $1.2 million for other improvements. One of the areas to be improved was the trail system. The Trails Discussion Group recommended changes to the park trail system. The physical changes suggested included eliminating unnecessary trails and minimizing new trail development, only mapping designated trails, and revegetating all other trails; adding needed signage indicating trail distance and destination as well as

Trails Plan Map

Trails Plan Map
Garden of the Gods Master Plan

exact current location; improving identification of Ute Trail; and placing signs to tell which uses are allowed on each trail.

Another area of concern was traffic in the park. The Traffic Discussion Group's goal was to help retain the pristine beauty of the Park. They recommended one-way roads with a bypass lane on steep grades; modifying or redesigning parking areas; providing improved on-street bicycling by including striped lanes or wide curb lanes on all paved roads, by implementing new bicycle parking sites throughout the park, and implementing a comprehensive bicycle information and regulation signage system with bike route signage; providing designated pullout observation areas with barriers to discourage their use as trailheads; providing adequate curbing and other features to prevent illegal parking; discouraging bike rental at the visitors' center, and not allowing bike rental in the park; and not allowing commercial, guided, or organized off-road bicycle trips in the park. In addition to these changes, all buildings with the exception of Hamp Hut, which is used by the Girl Scouts of the Wagon Wheel Council, and the Hidden Inn were to be torn down and their locations revegetated with native vegetation.[25]

In 1995, the work on the improvements in the Garden of the Gods received a boost in the form of a $100,000 grant from the Great Outdoors Colorado Trust Fund, which is funded by state lottery revenues. The Garden of the Gods grant was one of the largest awards in the state. The city was expected to match the grant with $383,000 from other sources. The money was used to redevelop two picnic areas, restore trails, and build auto pull-offs along the park's roads.[26]

As suggested in the master plan, a new sign was erected at the two entrances to the Garden of the Gods. Sign maker Tack Rice was commissioned to create the new signs. Rice combined technology and nature to carve seven rock slabs into three monoliths proclaiming the Garden of the Gods. The tallest sign stands eleven and a half feet tall and is twenty-seven feet wide. Completed in August 1995, the signs required months of planning and hunting in mountain quarries to hand-pick the rocks for the signs. The rock selected came from the Loukonen Brothers Stone Company quarry in Lyons, Colorado. The actual carving or sandblasting of the letters into the rocks was done after the slabs had been secured in their cement bases. While Rice was working on the rocks, some tourists asked him to move out of the way so they could get pictures. Rice, who majored in fine arts, said of the experience, "I always wanted to be an artist, but I never dreamed of doing it like this. This is a dream come true."[27] These new signs enhance the special atmosphere of the Garden of the Gods.

The Master Plan has been completed and implemented. Today, when a visitor walks through the Garden of the Gods, he finds vistas unimpaired by buildings or criss-crossed by roads and paths. The newly planted vegetation is taking hold and the area looks more natural than it has in the recent past. The more vulnerable areas are protected from the damage done by thousands of visitors' feet trampling on the delicate vegetation, or from being torn up by the actions of overly enthusiastic mountain bikers or horseback riders. Groups of visitors are led around the park by young people acting as park interpreters, helping them to understand the special nature of the Garden of the Gods. Thanks to the Master Plan, millions more visitors will be able to enjoy the special nature of the Garden of the Gods long into the future.

Endnotes:

1. *The Garden of the Gods Master Plan* (City of Colorado Springs Parks and Recreation Department, 1994), 12-13.
2. *Ibid.*, 12-14.
3. *Ibid.*, 13.
4. *Ibid.*, 13.
5. *Ibid.*, 14.
6. Barry Noreen, "Council to weigh Garden of Gods visitors' center," *Gazette Telegraph* (Colorado Springs: December 13, 1993).
7. *Ibid.*
8. Barry Noreen, "Private visitors' center OK'd for Garden of Gods: Council's vote unanimous," *Gazette Telegraph* (Colorado Springs: December 12, 1993).
9. *Ibid.*
10. *Ibid.*
11. *Ibid.*
12. Barry Noreen, "Indians want say at park," *Gazette Telegraph* (Colorado Springs: January 10, 1994).
13. *Ibid.*
14. Renaud, 17-18.
15. Barry Noreen, "Appeal filed to block park's visitors' center," *Gazette Telegraph* (Colorado Springs: January 21, 1994), A2.
16. Barry Noreen, "Visitors' Center foe rebuffed," *Gazette Telegraph* (Colorado Springs: Saturday, March 5, 1994), B1.
17. *Ibid.*
18. *Ibid.*
19. Barry Noreen, "Garden of Gods center plan unveiled," *Gazette Telegraph* (Colorado Springs: April 8, 1994).
20. Barry Noreen, "Future of park's Hidden Inn debated," *Gazette Telegraph* (Colorado Springs: April 6. 1994), B1, B8.
21. *Ibid.*, B8.
22. Barry Noreen, "Parks staff, residents debate fate of 79-year-old Hidden Inn," *Gazette Telegraph* (Colorado Springs: Friday, April 8, 1994), B2.
23. Collins, 17.
24. Pam Zubeck, "A wrecking ball is in Hidden Inn's future," *Gazette Telegraph* (Colorado Springs: Wednesday, August 27, 1997), News 1-2.

25. *Garden of the Gods Master Plan*, 43-46.
26. Barry Noreen, "Garden of the Gods work gets a lofty boost from trust fund," *Gazette Telegraph* (Colorado Springs: March 22, 1995).
27. Teresa Owen-Cooper, "Garden art on a grand scale," *Gazette Telegraph* (Colorado Springs: Sunday, September 17, 1995), B4.

The Future

That the Garden of the Gods is a special place is indicated by the great numbers of people who have come in contact with this geological wonder, and who have found it unique and often spiritual. Beginning with the geologic upheavals and subsequent erosions millions of years ago, nature began to set the stage for a long association with mankind. The spectacular red rock formations, the meeting place of three of the largest natural ecological zones, the wide variety of animal and plant life, and the healing qualities of the boiling springs at Manitou were further encouragement for the interaction of man with the Garden of the Gods. The current estimate of a million visitors a year to the Garden of the Gods in recent years is an indication of the popularity of this geologic wonder, and one can only speculate how many people have been impacted by the Garden of the Gods since prehistoric times.

As more archaeological explorations are conducted in the Garden of the Gods, we will, perhaps, move the earliest date of human contact with the Garden back even further than 1,300 B.C., as the oral traditions of the Ute people would indicate. We do not know if prehistoric peoples used the Garden of the Gods for spiritual purposes or simply found it an agreeable place to camp because of the abundance of game and plant life. Whichever the case, the Garden of the Gods has attracted people for thousands of years. The Ute people do consider the Garden of the Gods a place where the spirits dwell, and their ancestors camped in close proximity to the central garden area. They also used the boiling springs at Manitou for healing as well as religious purposes, as did other tribes who frequented the area.

With the coming of the Europeans, the Indians were forced out of this special place, and this is a point of conflict even today. The American Indian community is working to gain back their right to hold their annual Powwow in the Garden of the Gods once again, and to have the visitors to the park treat it with respect and to understand the full range of history of the Indians and the Garden of the Gods.

The early white explorers and settlers of the Pikes Peak area recognized the uniqueness of the Garden of the Gods, many writing about it in their letters and journals. As time passed, two individuals, William Jackson Palmer and Charles E. Perkins recognized the Garden of the Gods for what it is—a gift from God to all people. They wanted it to be open and free to everyone forever. Palmer and Perkins purchased as much of the land surrounding the central garden area as they could in order to protect this spectacular natural wonder from the inroads of development. In time, Charles Perkins' heirs donated the Garden of the Gods to the people of Colorado Springs and the world to use as a public park. Since that time, the city of Colorado Springs has purchased even more of the surrounding land to provide a buffer for the park from the encroachment of progress.

In modern times, the park has remained open to the public, though not always fully accessible as the Indians found when they tried to hold their Powwows there. The 1997 Master Plan was designed to return the Garden of the Gods as close as possible to its natural state, while still providing for a million or more visitors each year who come to the Garden of the Gods because it is special and unique. With proper management, millions of people from future generations will continue to find the Garden of the Gods a special and for many a spiritual place to visit.

About the Author

Toni Hamill, a retired teacher, resides in Monument, Colorado, where she taught English and American History at Lewis-Palmer High School. She graduated from Colorado Women's College (CWC) in 1966 with a BA in Liberal Arts and from the University of Colorado at Colorado Springs (UCCS) in 1997 with a Master of Arts degree in American Studies. While at UCCS she participated in an archaeological field school in the Garden of the Gods. The experience of excavating prehistoric sites inspired her to write her master's degree thesis about the people who have encountered the Garden of the Gods from prehistoric times to present. People of the Garden of the Gods is the result of that research.

Toni is married and has a son, 2 grandchildren, 4 bonus children and 10 bonus grandchildren. She is a member of the Zebulon Pike Chapter of the National Society Daughters of the American Revolution, the Mayflower Society, and various dog sport organizations.

When not writing, doing genealogical research, or having fun with her family, Toni trains and competes in dog agility and scent work with her 3 Australian Shepherds. She and her husband Bill enjoy traveling and taking active vacations around the world.

Bibliography

1900 Twelfth US Census, Topeka, Kansas.

"Act Out Fight For Suffrage." *The Rocky Mountain News* as reported in *The Colorado Prospector: Colorado History From Early Day Newspapers,* Vol. 17, No. 4, p. 6.

Arbogast, William R., Forest Tierson, and Alden Naranjo. *A Prehistoric Burial at 5EP2200, El Paso County, Colorado.* Colorado Springs: University of Colorado at Colorado Springs, 1996.

Beach, M. S. "El Paso County, the First Settlement." Colorado Springs Pioneers' Museum.

Belschner, Clara E. "Pioneer Tells of Ox-Cart Trip to C. S. Back in 1865." *Colorado Springs Sunday Gazette and Telegraph,* 3 November 1935, pp. 1-2.

Bird, Isabella L. *A Lady's Life in the Rocky Mountains.* 7th ed. New York: E. P. Dutton & Co., n.d.

Carey, David R. "1993 Garden of the Gods Archaeological Survey." Colorado Springs: City of Colorado Springs Parks and Recreation Department, 1993.

Cassells, E. Steve. *The Archaeology of Colorado.* Boulder: Johnson Books, 1983.

Cassidy, James J., Bryce Walker, Jill Maynard, and Judith Cressy. *Through Indian Eyes.* Pleasantville, New York: Reader's Digest Association, Inc., 1995. 300.

Chronic, Halka. *Roadside Geology of Colorado.* Missoula: Mountain Press Publishing Co., 1980.

Col, Jeananda. Zoom Dinosaurs. http://ZoomDinosaurs.com, 1996.

Collins, Michael H. "Historic Structures Report for the Hidden Inn, Garden of the Gods." City of Colorado Springs Parks and Recreation Department, 1996.

Colorado County Marriage Records and State Index, 1862-2006
"Colorado tourism's first family…," Jason Blevins, jblevins@denverpost.com, May 29, 2016.

Cragin Papers, XVIII. Colorado Springs Pioneers' Museum.

Cross, Charles. *Tourist Guide and Poetic Description of the Garden of the Gods, Glen Eyrie, Manitou and Vicinity*. Colorado Springs: February 7, 1906.

Ecology Center Proposed. *The Colorado Springs Gazette and Telegraph* as reported in *The Colorado Prospector: Colorado History From Early Day Newspapers,* Vol. 17, No. 4, p. 11.

Eighmy, Jeffrey L. *Colorado Plains Prehistoric Context. Denver*: State Historical Society of Colorado, 1984.

"Experts Differ Over Antiquity of Skeleton." *Colorado Springs Gazette,* Friday, 27 November 1925, p. 12.

Fetler, John. *The Pikes Peak People: The Story of America's Most Popular Mountain* Caldwell, Idaho: the Caxton Printers, Ltd., 1966.

Fremont, John Charles. "Report of the Exploring Expedition to the Rocky Mountains" *March of America Facsimile Series,* Number 79. Ann Arbor: University Microfilms, Inc., A Subsidiary of Xerox Corporation, 1966.

Garden of the Gods Folder. Colorado Springs Pioneer Museum.

Garden of the Gods Master Plan. City of Colorado Springs Parks and Recreation Department, 1994.

"Garden of the Gods Visitor Center Donates to City." *Senior Times of Colorado Springs, Colorado*. Vol. 5, No. 9, September 1996.

Gehling, Richard and Mary Ann. *Man in the Garden of the Gods*. Woodland Park: Mountain Automation Corp., 1991.

Gibney, Jim. "Generous Christmas Gift." *Denver Post* as reported in *The Colorado Prospector: Colorado History From Early Day Newspapers,* Vol.17, No. 4, p. 11.

Goss, Dr. James, ed. "Proceedings Garden of the Gods American Indian Workshop." Lubbock, Texas: Texas Tech University, 1994.

Guthrie, Mark R. Powys Gadd, Renee Johnson, Joseph J. Lischka. *Colorado Mountains Prehistoric Context*. Denver: State Historical Society of Colorado, 1984.

Hafen, Leroy R. *Ruxton of the Rockies*. Norman: University of Oklahoma Press, 1950.

_____. "Rufus B. Sage: Letters and Scenes in the Rocky Mountains," *Far West and Rockies Series, 1820-1875,* Vol. V (Glendale, California: Arthur H. Clark Co., 1956), 75.

_____. *The Mountain Men and the Fur Trade of the Far West, Vol. XI*. Glendale, California: Arthur H. Clark, Co., 1972.

_____. "Thomas Fitzpatrick and the First Indian Agency in Colorado," *Colorado Magazine*. Vol. VI (March, 1929): 53-62. State Historical Society of Colorado. 82-83.

Hartman, Todd. "Climbers Ignore Stone-Cold Fact." *Colorado Springs Gazette Telegraph,* 18 June 1997, p. 1A.

"Helen W. Stewart: Colorado's High-Altitude Hostess 1893-1979" *West Word*, Old Colorado City Historical Society. Vol 30, No. 6 August 2015.

"How Did Those Red Rocks Get There?" Garden of the Gods Geology Theater Program. Colorado Springs: City of Colorado Springs Parks and Recreation Department, 1993.

Howbert, Irving. *Indians of the Pikes Peak Region*. Glorietta, New Mexico: The Rio Grande Press, Inc., 1925.

_____. *Memories of a Lifetime in the Pikes Peak Region.*. Glorietta, New Mexico: The Rio Grande Press, Inc., 1970.

Hughes, David, "The Pastor's Bicycle and His Bible." Colorado Springs.yourhub.com, 8/22/2008.

Hughes, J. Donald. *American Indians in Colorado*. Boulder: Pruett Publishing Co., 1977.

"Indian Trail and Grave in Garden of Gods Now Marked By Huge Boulders." *Colorado Springs Sunday Gazette and Telegraph,* 15 February 1931, p. 6.

Ingersol, Ernest. *The Crest of the Continent: A Record of a Summer's Ramble in the Rocky Mountains and Beyond.* Chicago: R. R. Donnelley and Sons, Publishers, 1885.

"Introduction to the Ute Tribal History: Southern Ute Indian Tribe." Ignacio, Colorado: www.southern-ute.nsn.us/history/intro.html., 1997.

Jackson, Helen Maria Fiske Hunt. *Bits of Travel at Home.* Boston: Roberts Brothers, 1878.

Jones, Mary Kathryn. "Manitou Springs" *Colorado Country Life: The Official Publication of the Colorado Rural Electric Association*, Vol. 28, No. 6. Denver: Colorado Rural Electric Association, June 1997. 17.

Kehoe, Alice Beck. *North American Indians: A Comprehensive Account.* Englewood Cliffs: Prentice Hall, Inc., 1981.

Kennedy, Roger G. ed. *The Smithsonian Guide to Historic America: The Rocky Mountain States.* New York: Stewart, Tabori, and Chang, 1989.

Kim & Pat Messier's Blog, "Strausenback as Artist. Part II: Lithographs and Etchings," May 30, 2022."

Kim & Pat Messier's Blog," Strausenbacks as Artist, Part I: Original Art, January 2, 2023.

Liller, J. E. *El Paso County*. University Place, Nebraska: Charles Elce and Son, 1924.

Lindbergh, General Charles A. Letter to the Garden of the Gods Preservation Council,

24 August 1968. Garden of the Gods Folder, Colorado Springs Pioneers' Museum.

"Lost Cave in Garden of Gods Found and Sealed Up; Earliest of Pioneers Carved Names and Dates There." *Colorado Springs Sunday Gazette and Telegraph,* 13 October, 1935, p. 4.

Ludlow, Fitz Hugh. *The Heart of the Continent: a record of travel across the plains and in Oregon, with an examination of the Mormon principle.* New York: Hurd and Houghton, 1870.

Maxwell, James A. ed. *America's Fascinating Indian Heritage.* Pleasantville, New York: Reader's Digest Association Inc., 1978.

McConnell, Virginia. "Real Indians Highlighted Shan Kive: Festival Predecessor to the Rodeo." *The Free Press.* Colorado Springs: (196?): 4-5.

Marsh, Charles S. *People of the Shining Mountains.* Boulder: Pruett Publishing Co., 1926.

Messier, Kim and Pat**,** *Images of America, Garden of the Gods Trading Post.* Charleston; Arcadia Publishing, p. 18.

Messier, Pat and Kim, 2014 Reassessing Hallmarks of Native Southwest Jewelry. Schiffer Publishing pp 41-45.

Missouri Democrat, 20 November 1859. Colorado Historical Society, MSSXXV, p. 56.

Naranjo, Alden. E-mail interview by author,. Monument, Colorado, 8 April 1997.

Nesbit, Paul A. *Garden of God's.* Colorado Springs: Paul A. Nesbit, 1968.

"New Indian Store in Garden of the Gods.", *Colorado Springs Gazette,* April 7, 1929.

"New Mexico: The Onate Expeditions," www.wisconsinhistory.org, 10/2/2008.

"New Roads in Garden Gods?" *Colorado Springs Gazette and Telegraph,* 3 November 1935, p. 1.

Newcomb, William W. Jr. *North American Indians: An Anthropological Perspective.* Pacific Palisades: Goodyear Publishing Company Inc., 1974.

Noreen, Barry. "Appeal filed to block park's visitor's center." *Colorado Springs Gazette Telegraph,* 21 January 1994, p. A2.

_____. "Council to weigh Garden of Gods visitors' center." *Colorado Springs Gazette Telegraph,* 13 December 1993.

_____. "Future of park's Hidden Inn debated." *Colorado Springs Gazette Telegraph,* 6 April 1994, pp. B1, B8.

_____. "Garden of Gods center plan unveiled." *Colorado Springs Gazette Telegraph,* 8

April 1994.

_____. "Garden of the Gods work gets a lofty boost from trust fund" *Colorado Springs Gazette Telegraph,* 22 March 1995.

_____. "Indians want say at park." *Colorado Springs Gazette Telegraph,* 10 January 1994.

_____. "Mountain biking ban to be enforced." *Colorado Springs Gazette Telegraph,* 3 July 1993.

_____. "Parks staff, residents debate fate of 79-year-old Hidden Inn." *Colorado Springs Gazette Telegraph*, 8 April 1994, B2.

_____. "Private visitors' center OK'd for Garden of Gods: Council's vote unanimous." *Colorado Springs Gazette Telegraph,* 12 December 1993.

_____. "Visitors' Center few rebuffed." *Colorado Springs Gazette Telegraph,* 5 March 1994, B1.

Ormes, Manly Dayton. *The Book of Colorado Springs.* Colorado Springs: The Dentan Printing Co., 1933.

Owen-Cooper, Teresa. "Garden art on a grand scale." *Colorado Springs Gazette Telegraph,* 17 September 1995, B4.

Perkins, C. E. "Dear Sir." William Jackson Palmer Papers, Colorado Historical Society, Denver, Colorado. *Colorado Prospector,* Vol. 17, No. 4, p. 3.

Pettit, Jan. *Utes The Mountain People.* The Revised Edition. Boulder: Johnson Books, 1990.

Pike, Zebulon M. *The Southwestern Expedition of Zebulon M. Pike.* Chicago: Lakeside Press, 1925.

Pomeroy, Earl Spencer. *In Search of the Golden West; the tourist in western America.* New York: Alfred A. Knopf, 1957.

"Professor Kerr finds fossils – *Scientific American.*" Colorado Springs Gazette, 23 February, 1878, p 4.

"Rare Honey." *The Colorado Free Press,* 20 April 1962, as reported in *The Colorado Prospector: Colorado History From Early Day Newspapers,* Vol. 17, No. 4.

Reed, Bill. "Fossil Fuels New Discovery: Scientists name species after Garden of the Gods." *The Gazette,* 25 May, 2008, Life 1, 10.

Renaud, E. B. "Western and Southwestern Indian Skulls." Anthropological Series, First Paper. Denver: University of Denver Department of anthropology, 1941.

Richardson, Albert D. *Beyond the Mississippi: From the Great River to the Great Ocean;*

Life and Adventure on the Prairies, Mountains, and Pacific Coast, 1857-1867. Hartford: American Publishing Co., 1867.

Ross, Thomas E. and Tyrel G. Moore, ed. *A Cultural Geography of North American Indians.* Boulder: Westview Press, 1987.

Sage, Rufus B. "Letters and Scenes in the Rocky Mountains," in *Far West and Rockies Series, 1820-1875, Vol. 5.* edited by LeRoy Hafen. Glendale, California: Arthur H. Clark Co., 1956.

"Skeleton Found In Garden of the Gods," *The Colorado Springs Gazette,* 2 November 1925, as reported in *The Colorado Prospector: Colorado History From Early Day Newspapers,* Vol. 17, No 4, pp. 6-7.

"Skeleton Was Indian Woman, Experts Aver." *Colorado Springs Gazette*, Friday, 6 November, 1925, p. 9.

Skodack, Debra. "Guardian of the Garden." *Colorado Springs Gazette Telegraph,* Sunday, 13 July 1980, 5BB.

Smith, Eugene R. *Mystery Man.* Denver: State Historical Society of Colorado, 1996.

Sons of Dewitt Colony Texas, *New Spain-Index* (McKeehan, Wallace L., 1997-2007).

Sprague, Marshall. *Newport in the Rockies.* Denver: Sage Books, 1961.

Stone, Wilbur Fisk, Ed. *History of Colorado*, Volume 1. Chicago: The S.J. Clarke Publishing Co., 1918.

"Strausenback Wins Renown as Artist." Transcription from *Colorado Springs Gazette,* Jan 8, 1936, p. 12.

Street, Julian. *Abroad at Home: American ramblings, observations and adventures.* Garden City, New York: Garden City Publishing Co., 1914.

Trading Post. Square Space.com/history. Accessed 7/24/23.

Ubblohde, Carl, Ed. *A Colorado Reader*. Boulder: Pruett Press, Inc. 1964.

"Uto-Aztecan Languages." (http://en.wikipedia.org/wiki/Uto-Azetecan_Languages. 2008), 4-5.

Walker, Melissa. "Clues to Prehistoric Life and Landscapes of Garden of the Gods Park." (Colorado Springs, Colorado) 1998.

_____. "Dwelling Among the Wonders and Mysteries of Life." *The Bulletin: Summer '98:* (Colorado College Publications), 1998.

Wallace, Ernest and E. Adamson Hoebel. *The Comanches, Lords of the South Plains.* Norman: University of Oklahoma Press, 1952.

Weber, David J. *The Taos Trappers: The Fur Trade in the Far southwest, 1540-1846.* Norman: University of Oklahoma Press, 1971.

Wilcox, Ted. "Garden of Gods Keeps Message 37 Years." *Colorado Springs Gazette Telegraph,* 1961.

Woodard, Bruce A. *The Garden of the Gods Story.* Colorado Springs: Democrat Publishing Co., 1955.

Zubeck, Pam. "A wrecking ball is in Hidden Inn's future." *Colorado Springs Gazette Telegraph,* 27 August 1997, News pp. 1-2.

Index

A

Adams-Onis Treaty 55
American School for Research and Museum at Santa Fe 50
Apache 16, 21, 22, 28, 35, 36, 38
Arapaho 21, 28, 29, 35, 36, 39, 40
Archaic People 17
Atlatl 15

B

Balanced Rock 3, 5, 6, 87, 93, 106, 118, 133
Beach, Melancthon S. 73, 80-82, 89, 94, 95
Bell, Captain John R. 57
Bemis-Taylor Foundation 132
Bell, William C. 98
Bent, William and Charles 55, 56, 59
Bent's Fort 59
Bijeau, Mr. 57, 58
Bird, Isabella 11, 101, 102
Blair Athol 11, 29
Brill, Kathleen 134
Buckskin Charley 119

C

Cable, Rufus E. 73, 80-82, 94, 95
Capota 30
Carle, William F. 127
Carpenter, Dr. Ken 133, 134
Carson, Kit 59, 61
Carver, Jonathan 55
Ceramic or Woodland Culture 17
Chambers Ranch 92, 110
Cheyenne 2, 21, 28-30, 35, 39, 40, 50, 62, 91, 93, 105
Chipeta 94, 119, 120, 124
Civilian Conservation Corps 67, 125
Colorado City 45, 47, 67, 69, 73, 74, 79-83, 88-97, 100, 102, 105, 110, 111, 118, 119, 123, 125, 127, 134
Colorado City Town Company 73, 80, 90

Colorado Enabling Act 88
Colorado Piedmont 3
Colorado Springs Company 96, 98
Colorado Springs Jaycee's 129
Colorado Woodland People 17
Colorow 93
Comanche 16, 21, 28, 29, 36, 37, 38, 39, 50, 51

D

Davis, R.S. 126
de Onate, Juan 53
de Villasur, General Pedro 54
de Zaldivar, Juan 53
Denver City Company 73
Denver Museum of Natural History 133, 134
Dickson, T. C. 69, 70
Dodge, Colonel Henry 59

E

El Paso City 80
El Paso County Pioneer Society 78, 89
El Pomar Foundation 132

F

Fatty Rice 84, 105, 106
Fitzpatrick, Thomas 62
Fontaine qui Bouille 9, 46, 59, 79
Fountain Creek 2, 5, 9, 45, 55, 62, 79, 88, 93
Fountain Formation 1, 5
Fremont, John C. 59
Friends of Garden of the Gods 138
Frost, Louisa B. 76

G

Galloway, Walter 95
Garden of the Gods Trading Post 107, 109, 139
Garvin, Bill 95
Gateway Rocks 5, 8, 86, 104-106, 129
Giant Footprints 5
Gilpin, Governor William 82, 89
Glen Eyrie ii, 2, 11, 12, 29, 33, 79, 82, 91, 97, 101-105, 112
Goerke, Curt 106, 118
Goss, Dr. Jim 23

Grove, Bertha 33

H

Haas, Terry 139
Hall, James 60
Hartley, William 69, 70
Hidden Inn 125-127, 131, 134, 135, 141-145
Hewett, Dr. Edgar L. 49, 50
Howbert, Irving 47, 54, 82, 88, 91-93, 98, 121
H. W. Stewart Inc. 126

I

Ingersoll, Ernest 9, 102
Iron Ute Spring 9, 10

J

James, Dr. Edwin 46, 58, 59
Jaycee 129, 130
Jewett, Marshall M. 73, 74
Jimmy Camp 69, 70

K

Kerr, James Huchison 101
Kiowa 28, 29, 35, 36, 93
Kroenig, William 56

L

Lakota Sioux 40
Larimer Party 70, 73
Lawrence Party 68-71
Ledoux, Abraham 57
Lindbergh, Charles 130, 131
Little Chief Spring 9
Long, Colonel Stephen 46
Long, Major Stephen H. 57, 58, 59
Long Expedition 58
Luce, Reverend Albert W. 127, 128
Lyons Fault 8
Lyons Sandstone 1

M

Manitou Springs ii, 5, 6, 8-12, 33, 36, 40, 59, 86-88, 91, 96, 97, 99, 106, 111, 120, 122, 126, 131, 138

Marsh, Dr. O.C. 101
Master Plan 9, 19, 127, 137, 138, 140, 143, 144, 147, 150
McDowell, J.L. 87
Mouache 30

N

Naranjo, Alden 21, 23, 25, 27, 30, 33, 51, 151
National Woman's Party 124
Native Hearts Council 140
N.A.T.I.V.E.S. 133, 143, 144
Navajo 9, 22, 46, 106, 107, 108
Nuche 30

O

Ouray 92, 93, 94, 120

P

Pabor, William E. 98, 99
Palmer, Queen 98
Palmer, William J. 32, 33, 58, 62, 67, 96-98, 101-105, 110-112, 132, 150
Palmer Park Civilian Conservation Corps 67
Pawnee 21, 28, 39, 40, 50, 57, 68, 76
Perkins, Charles E. 104, 105, 111, 112, 115-118, 125, 126, 138, 150
Pierre Shale 2
Pike, Zebulon 56, 57, 59
Pikes Peak or Bust Rodeo 124
Pleistocene Ice Age 3

R

Rampart Range Fault 2
Red Rock Canyon 2, 11, 29
Renaud, E.B. 50, 52
Rice, Fatty 84, 105, 106
Richardson, Albert D. 102
Richter, Stuart W. 132
Rock Ledge Ranch Historic Site 110
Rocky Mountain Brewing Company 74
Ruxton, Lt. George Frederick 9, 10, 45-47, 61, 62, 127

S

Sage, Rufus B. 46, 59
Seven Minute Spring 10
Shan Kive 121

Shoshone 9, 21, 28, 36, 37, 38, 39
Sierra Club 139, 142
Sleeping Giant 5, 8
Soda Springs 56, 88, 95, 96, 97, 98, 123
Spain 53, 55, 56,
Spaulding, Jacob 62
St. Vrain, Ceran 56
State Historical and Natural History Society of Denver 49
Stewart, Helen Wilson 126, 127
Stewart, Orrie 127
Stratton Spring 10
Strausenback, Charles E. 106 -109, 126
Sulfur Spring 9

T

Tabewache 30
Tappan, Lewis N. 81, 95
Terryall District 80
Toad and Toadstools 6
Tower of Babel 5, 8
Trading Post 107, 109, 126, 139
Twin Spring 10

U

University of Denver 50, 156
Ute 2, 8, 9, 10, 13, 20-23, 27-31, 33, 34, 39, 40, 45, 48, 49, 51, 55-57, 59, 62, 68, 73, 80, 88, 89, 92-94, 96, 101, 102, 119, 121-124, 132, 144, 149
Ute Pass Fault 2, 8
Ute Soda 9

W

Walker, Melissa 133, 136
Wheeler Spring 10
White House 110, 132, 138, 139
White House Ranch Living History Association 138, 139
White Rock 5, 65, 67, 68, 75, 77, 79, 104
Wooton, Dick 56, 95
Wright, Andrew C. 68

Z

Zang, Philip 74

www.ingramcontent.com/pod-product-compliance
Lightning Source LLC
Chambersburg PA
CBHW061208230426
43665CB00028B/2952